Wearable Solar Cell Systems

Wearable Solar Cell Systems

Authored by

DENISE WILSON

CRC Press
Taylor & Francis Group
Boca Raton London New York

CRC Press is an imprint of the
Taylor & Francis Group, an **informa** business

CRC Press
Taylor & Francis Group
6000 Broken Sound Parkway NW, Suite 300
Boca Raton, FL 33487-2742

First issued in paperback 2022

ISBN-13: 978-0-367-02347-8 (hbk)
ISBN-13: 978-1-03-233756-2 (pbk)
DOI: 10.1201/9780429399596

Publisher's Note

The publisher has gone to great lengths to ensure the quality of this reprint but points out that some imperfections in the original copies may be apparent.

Library of Congress Cataloging-in-Publication Data

Names: Wilson, Denise (Electrical engineering professor), author.
Title: Wearable solar cell systems / by Denise Wilson.
Description: First edition. | Boca Raton, FL : CRC Press/Taylor & Francis
  Group [2020] | Includes bibliographical references and index.
Identifiers: LCCN 2019037689 (print) | LCCN 2019037690 (ebook) | ISBN
  9780367023478 (hardback) | ISBN 9780429399596 (ebook)
Subjects: LCSH: Solar cells. | Wearable technology--Power supply.
Classification: LCC TK2960 .W55 2020 (print) | LCC TK2960 (ebook) | DDC
  621.31/244--dc23
LC record available at https://lccn.loc.gov/2019037689
LC ebook record available at https://lccn.loc.gov/2019037690

Visit the Taylor & Francis Web site at
http://www.taylorandfrancis.com

and the CRC Press Web site at
http://www.crcpress.com

To the hope for the planet that renewables

provide… not the least of which is hope itself.

Contents

List of Figures and Tables

Preface

As much as we use and rely on myriad portable and wearable devices to carry out our daily lives, they consume surprisingly little power. That is—compared to the sheer amount of overall electricity we consume in modern society every day. That is—if we ignore the more invisible power that is consumed in the work of towers in the cellular network, the machinations of the cloud, and the manufacturing, transport, and distribution processes that go into bringing these devices to our doorstep. That is—if we ignore all the energy that goes into extracting and processing the natural resources that feed our ever increasing ravenous appetite for batteries and other components that are integral to our expanding repertoire of electronic gadgets.

Even so, the impact of our portables and wearables—in size, shape, weight, and footprint—still appear to be insignificant compared to the rest of all things electronic. Given this apparent insignificance, why expend the natural resources, energy, and effort to write a book on alternative means to power these devices? Why not continue the focus on alternative and renewable energy dedicated to providing power to residential homes, to industrial operations, to the increasing numbers of electric vehicles rolling around the planet?

There are many answers to these questions. Philosophically, these questions and their many answers make for a book by itself. Not being a philosopher and restricting myself to science and engineering, however, I can only offer a few thoughts in equally few paragraphs.

As the number and functionality of portables and wearables continue to grow, they will not only consume more power but will also continue to displace more stationary electronics. How many of us now watch streaming entertainment on our smartphones rather than movies and shows on an actual television? How many of us now use our smartphones to check e-mail rather than using a desktop computer to do so? As we lean more and more on the convenience of portable devices, these devices will naturally expand in their functionality and, not surprisingly, demand more power to do what they do.

As the markets for portables and wearables grow, they feed an imagination and ambition for more portables and more wearables to do even more things. If the power constraint were to disappear, if the need to tether to a wall outlet for recharging were to vaporize, if the weight of the portable power bank were to stabilize rather than increase alongside the number and capacity of our gadgets, what could then be possible in what we wear and carry? In healthcare, entertainment, industrial process control, occupational and personal safety, and many more applications—the possibilities seem endless.

But, solar energy is not the only source of renewable energy available to the individual in everyday activities. Is solar energy to power wearables and portables a viable or a laughable option? As individuals spend more and more time indoors, the human body's access to sunshine continues to dwindle. As individuals continue to be on the move, harvesting energy from their motion seems to be a more reliable way to provide energy to the devices they carry. Yet, even as we become more and more creatures of living, working, and playing indoors, we are not simultaneously turning the lights off. While not as strong or powerful as the sun, these lights, whether in the office, the living room, or the classroom, nevertheless provide a continuous, reliable, and persistent source of energy to wearable solar energy systems. And, while developed to process sunshine rather than artificial light, existing solar cells and state-of-the-art knowledge regarding how they work and how to design them into systems provide an excellent foundation for wearable solar cell system technology. Capitalizing on the inertia of traditional solar systems gives wearable solar a leg up over other means of harvesting energy, including converting the body's mechanical energy or heat energy to electricity.

Given the potential, abundance, and sustainability of solar and light energy, harnessing it to meet our ravenous appetite for electricity has attracted a lot of attention. The number of manuscripts, articles, and books that focus on photovoltaic-based solar energy systems is impressive, and so it should be given the enormous dedication the scientific community has shown to develop more efficient means to convert the sun's energy to electrical energy. As unlikely as the indirect bandgap of silicon might be to support an empire of solar panels, that is exactly the story that the solar panels, modules, and systems based on crystalline silicon tells in the overwhelming dominance of these technologies in global market share. Such dominance has only come about with both a powerhouse of scientific talent and a societal commitment to clean and renewable energy. Seeking not only market share but a cure to the remaining ailments of silicon solar cells, second and third generations of photovoltaic technologies represent a remarkable imagination and collective understanding in the scientific community that merits further publication and dissemination.

Despite the wealth of knowledge and insight of both science and engineering into how photovoltaic solar cells work and solar system designs can be optimized, there is still work to do in understanding how these technologies can be adapted to power portables and wearables. New design challenges emerge and design priorities change as the harvesting of light energy is adapted from the rooftop to the human body. Unlike many manuscripts that contextualize photovoltaics around the rooftop, this text explores the potential and capacity of the human body in everyday activity to exploit, harness, and harvest light energy in a multitude of situations in order to power our increasing appetite for mobile electronics—both wearable and portable.

With the rooftop in mind but the body in its crosshairs, this book explores the multiple generations of photovoltaic solar cell technologies and the many possibilities of applying them to portable and wearable energy harvesting.

As I have written this book over many months, I have been reminded over and over again that, as many an author and writer can attest, writing a book is no minor pastime. As this book has evolved over the past year, I have many people to thank for their support and encouragement. I am truly grateful for my two favorite readers, my husband Barry and sister Heidi, for reviewing, editing, and reviewing again a seemingly unending stream of drafts.

I am grateful for the students in my classes who have continued to challenge me over the years to find different explanations and alternative ways to explain science and engineering in ways that make it easier for minds to grow and master difficult concepts. Without the questions, the office hours, and interaction with so many different ways of thinking, I might find myself writing to an audience of only one.

I appreciate the support, grace, and understanding of the individuals at CRC press who have remained alongside me in preparing and publishing this manuscript: Nora Konopka, Prachi Mishra, Revathi Viswanathan, and Michele Smith. Without their patience, my experience in writing this manuscript would have been truly painful.

And, I owe a debt of gratitude to the many others who have stood by me through the writing process, especially Kelley Miles who has adopted the book chicken as her own and sent the chicken back into my life when I might have otherwise have sent it to the chopping block. Said chicken ran around my house, my car, my office, squawking ever more loudly as the manuscript deadline approached with a very special cluck that sounded suspiciously like "book, book, book, book" rather than the traditional squawk of more normal chickens.

And, finally, my deepest gratitude to the Creator of all this fascinating science upon which our many technologies rest. Without the gifts He has provided me, my writing on the topic of solar cells would sound much more like a clucking chicken than anything close to articulate prose.

About the Author

Denise Wilson is a professor in the Department of Electrical and Computer Engineering at the University of Washington in Seattle where she has worked since 1999. Previously, she held a similar position at the University of Kentucky in Lexington, Kentucky.

Denise is also founder and managing director of Coming Alongside, an environmental services non-profit organization whose mission is to make hazards posed by the environment to human and animal health visible and actionable. She received a BS degree in mechanical engineering from Stanford University (1988); MS and PhD degrees in electrical engineering from the Georgia Institute of Technology, Atlanta, in 1989 and 1995, respectively; and an MEd degree from the University of Washington in 2008.

Denise has published over 40 articles in peer-reviewed journals and over 100 articles in peer-reviewed conferences on topics ranging from circuit design to environmental health. She has also published two books and three book chapters and developed extensive web-based educational materials in educational research, environmental health, and the environmental impacts of technology.

Denise has taught a wide range of university-level courses at both undergraduate and graduate levels on topics related to the environmental and social impacts of technology, sustainable design for the developing world, impacts of natural disasters, circuits, sensors, and semiconductor devices. Her research focuses on both engineering education as well as photovoltaic and sensor systems with specific interests in applying these devices to solving problems in environmental monitoring.

1

The Power of Light

Smartwatch? Fitness tracker? Portable electrocardiogram (ECG)? Smartphone? Posture monitor? Hearing aid? MP3 player? e-reader? Wireless headset? Hiking watch? Gaming headset? Sleep monitor? Laptop computer? Movement assistance? Portable power bank? Tablet? Heads-up display?

How many of these devices do you already have? How many do you not yet have but still want? How many are you willing to wear before you say enough is enough? These may be questions to think about only on Amazon Prime Day when a wide range of tantalizing electronic devices appears before you at bargain prices. Or, as an individual on the move much of the time, you may have found that many of these devices increase your productivity, support good health, improve quality of life, or simply make life a little easier.

But is a portable power bank enough to power this plethora of devices in an increasingly mobile and fast-paced society? And if a power bank is not enough, where will the energy come from? As you look at the solar panels on a nearby roof, do you wonder if there is enough light in your daily travels to support powering the sum total of electronics you carry on your person using solar panels worn where else—but on your person?

Is the Power of Light Enough?

On the one hand, there is no shortage of people who will laugh at the idea of installing rooftop solar panels on residential homes in a place like Seattle where the sun shines only 152 days a year on average (Best Places n.d.), it rains 155 days a year, and on the winter solstice, daylight is barely longer than 8 hours in a day (Clarridge 2016). Yet, despite the dismal gray weather that plagues the Emerald City most of the year, Seattle is a solar leader among cities at 25.7 watts (W) of installed solar photovoltaic (PV) power per person and a total solar-generated capacity of 18.1 MW (Bradford and Fanshaw 2018). How can this be possible?

As is the case with many technologies, the goodness of the technology cannot be distilled to a single figure. It is true that Seattle does not get the best sunshine in the world—ranging from 0.92 kWh/m²/day in December to 5.88 kWh/m²/day in July compared to Phoenix, Arizona, which ranges from 2.95 kWh/m²/day in December to 7.52 kWh/m²/day in June (Boxwell 2019). But, how much of a difference does this make in the feasibility of powering a home on solar energy in these two locations. Arizona has a per household electricity usage of 1,034 kWh/month, about 35 kWh/day, and 12,408 kWh/year (Crees n.d.). Providing this level of energy through purely solar means would require a maximum of about 74 m² of solar panels on the rooftop, assuming solar panels that work at 16% efficiency (NREL n.d.). This represents less than a third of a residential rooftop with a surface area of 241 m² based on an average 2,600 sq. ft. home with a garage (Center for Sustainable Systems n.d.). With its far less sunny climate, Washington state has a per household electricity usage of 955 kWh/month, about 32 kWh/day, and 11,460 kWh/year (Crees n.d.) which requires a maximum of about 217 m² of solar panels—about 90% of rooftop area in an average home (Center for Sustainable Systems n.d.). These two examples suggest that commercially available solar panels and the mature technology that these products now use can indeed power homes in both sunny climates and more dreary climates. In addition to basic facts regarding the availability and cost of solar energy including subsidies, policy and politics also go a long way toward explaining why some cities surge ahead in solar capacity, while others lag behind. In the long term, as the price of solar panels comes down and technology advances to allow for more than the flat, rigid surfaces of a roof to be covered with solar cells and systems, the idea of powering a home solely on solar energy becomes far less laughable (even during the short, rainy, gloomy days of winter in the Pacific Northwest).

Similar to solar energy in the Pacific Northwest, solar systems worn on the body may at first appear laughable, absurd, or misguided. After all, people in developed societies spend an increasing amount of time indoors (Walden 2018) where access to light energy is far lower than it is in the outdoors during daylight hours (Littlefair 1985; Ledke Technology Ltd. n.d.). Lower light intensities combined with the low energy demand of portable and wearable devices relative to overall electricity consumption should make investments in wearable solar cell systems take a backseat to advancements in traditional stationary solar systems. For the increasing numbers of nomophobes (Yildirim and Correia 2015) who have genuine separation anxiety when their phones are somewhere they are not, portable or wearable solar cells may be absolutely critical for a backcountry hike or other life activity that is disconnected from alternating current (AC) power. Otherwise, portable battery banks should be able to meet the expectations of the average gadget-loaded individual. After all, most individuals can simply wait until they can gain access to an AC or other recharge station to recover access to their myriad of portable and wearable devices.

Or not.

1.1 Portable, Mobile, and Wearable Devices

Distinctions among portable, mobile, and wearable devices used to be clear. A portable device was an electronic device that could be carried around relatively easily—lightweight and not too bulky. A mobile device referred to a portable device that allowed its users to be mobile, typically through a meaningful degree of connectivity, whether Wi-Fi or cellular. A wearable device was one that did not need to be carried but managed to stay on an individual's body without assistance, albeit with no connectivity to the rest of the world.

As devices have evolved to have more and more functionality, these distinctions have rapidly faded. For example, a smartwatch remains on the wrist without any assistance from the user (making it wearable), but the watch also provides a level of connectivity which allows individuals to be disconnected from wired computing (making the same watch also mobile). A laptop computer, in the absence of Wi-Fi (or a personal hotspot) is portable but not mobile, but can quickly become mobile when it sniffs out an accessible Wi-Fi connection.

While the distinction among portable, mobile, and wearable devices continues to get fuzzier, the sales of these devices continue to climb. In 2018, over 160 million notebook computers (Holst 2018), 172 million wearable devices (International Data Corporation 2019), 173.8 million tablets (Barbaschow 2019), and 1.43 billion smartphones (Barbaschow 2019) were sold around the world. For the most part, notebook/laptop and tablet computers consumed the most power, followed by smartphones, with most wearable devices consuming the least power. One noteworthy exception among wearable devices is a virtual reality headset that can consume even more power than laptop computers and has a battery life of only 2–3 hours. In contrast, some wearable devices like hearing aids can consume less than a percent of what laptop and tablet computers consume and have a battery life measured in days or weeks.

Given the ambiguity of terms among portable, wearable, and mobile devices, consistent terminology will be used throughout this book. The term portables will be used to describe general-purpose computing devices that can be readily carried by most individuals including laptop, notebook, and tablet computers as well as smartphones, traditional flip phones, e-readers, and similar devices. In contrast, the term wearables will be used to describe most of the rest of the devices that can be worn on head, arms, ears, wrist, feet, back, abdomen, and other parts of the body or clothing where they do not require user intervention to keep them there.

A vast majority of portables and wearables are powered by lithium-based batteries with the exception of very low-power wearables (e.g., many hearing aids) that continue to employ single-use batteries. Most of these lithium-based batteries are rechargeable and rely on traditional AC wall outlets or Universal Serial Bus (USB) chargers that can use direct current (DC) for recharging. While a small percentage of battery recharging cycles can be traced back to renewable energy sources such as wind and sun, most trace back to the coal and natural gas that remain responsible for producing most of the electricity in the world. Given the importance to many of never having a battery fail while on the move and the increasing number of portables and wearables that run that risk, wearable sources of electricity production are growing in popularity. Moreover, the electricity (and recharge capacity) associated with most of these wearable sources will depend, to some extent, on an individual's activity. The harvesting of solar and light energy for wearable, portable, and electronic devices is no exception. Given the small amount of light energy available for harvesting via a wearable solar cell system compared to traditional solar energy systems, the potential impact of wearable systems can at first glance seem wildly insignificant. But, a closer look can reveal otherwise.

1.2 Impacts of Wearable Solar Cell Systems

1.2.1 A Mere Drop in the Energy Bucket

In New York City, a smartphone is charged 2.7 times per person per day on average (Veloxity 2017). Sounds like a lot? Maybe. 2.7 times a day is a lot in behavioral terms, but in energy terms, it's not so much. Consider a typical iPhone that has a battery capacity of about 2,500 mAh (SocialCompare 2019) and operates at about 3.7 V. A full charge of the battery consumes 9.25 Wh of energy or 0.00925 kWh. Thus, 2.7 charges a day consume 0.025 kWh/day and 9.12 kWh/year—which amounts to approximately 0.071% of the average 12,900 kWh of electricity consumed per person per year in the United States (The World Bank 2014). A mere drop in the bucket.

For an individual in Kenya consuming 164 kWh/year, the same level of phone usage amounts to about 5.6% of the 164 kWh of electricity consumed per person per year (The World Bank 2014). A bigger drop in the bucket, but still a drop. In the big picture, many other devices and appliances consume far more energy than a single portable or wearable device. And, if electricity or energy usage were all that mattered in the powering of these devices, there would be little reason to pursue renewables to power the portables and the wearables in the world.

1.2.2 An Issue of Convenience

While the electricity consumed by portables and wearables may seem trivial, their energy demand is compounded by access to available power. A large majority of these devices are powered by rechargeable batteries that require AC (wall outlet) power to recharge. Convenient access to AC power no longer keeps up with the battery usage for many individuals. The magnitude of this inconvenience and the extent to which consumers will go to avoid sudden battery death, particularly with smartphones, is reflected in the rapidly growing portable battery pack market. The market for these portable power banks is expected to grow to over 10 billion dollars by 2020, with a compound annual growth rate of 17.5% between 2014 and 2020 (Markets and Markets 2014).

And, weighing in at a little more than half of a pound for up to 10,000 mAh of charging capacity (Power Bank Guide 2019), these power banks seem more than adequate to resolve issues of convenience for portable and wearable device users. However, as the number of devices that an individual carries increases, so does the total device weight and so does the weight of a portable power bank that can keep those devices charged successfully. In the absence of a leap forward in battery capacity, keeping up with consumer expectations of convenient and accessible charging by relying on portable power banks may be untenable.

1.2.3 Circling Back to the Environment

The push for widespread renewable energy is driven in part by the continued depletion of fossil fuel reserves and also by the significant environmental impacts of these nonrenewable fuels. Electricity derived from solar or light energy is considered to be one of the cleanest and most abundant renewable resources. Such renewable resources are an essential part of constraining global temperature increases to less than 2°C in order to avoid catastrophic impacts on the planet (McSweeney, Pearce, and Prater 2018). More renewable energy is only part of the solution to achieve the 2C limit. Greenhouse gas emissions associated with fossil fuel use will also have to decrease significantly using novel, cleaner methods. Considering advances in fossil fuel-based electricity and renewable energy production that are consistent with a 2C limit, fossil fuels will still outpace renewables in overall emissions. By 2050, it is expected that solar energy will produce only 6 g of carbon dioxide equivalent (i.e., greenhouse gases) per kilowatt-hour of electricity generated, while coal will still produce 109 g of carbon dioxide equivalent gases (Evans 2017). In addition to vastly reduced greenhouse gas emissions, PV-based solar energy also consumes far less water during production and use compared to the large amounts of water used when coal is burned to generate electricity. Furthermore, while solar panels consume surface area, they consume no usable land when installed on rooftops or in wearables. And, along with no emissions during use, their impacts on both air and water pollution are minimal compared to other means for producing electricity. For its inherent environmentally friendly nature, as the costs of residential solar installations, both standalone and grid-connected, continue to decrease, solar energy has a firm foothold in the future of renewables. But, given the small fraction of overall electricity consumption that portables and wearables consume, the environmental impacts of putting solar energy to work in wearable solar cell systems seem to be quite small.

However, the displacement of fossil fuels through wearable energy harvesting systems is only one piece of the environmental impact puzzle. At present, wearables and portables rely heavily on rechargeable batteries based on lithium and single-use batteries based on alkaline, silver oxide, lithium, and zinc air chemistries. The demand for these batteries is skyrocketing as the number of portables and wearables proliferates. The environmental impact of supporting this uptick in battery demand is profound. Billions of single-use batteries are disposed of every year. A majority of these single-use batteries are alkaline that some argue can cost more in environmental terms to recycle than to leave in a landfill. But, placing these billions of batteries in landfills also has more impact on marine, freshwater, and terrestrial toxicity than recycling them (Xará, Almeida, and Costa 2015), and certainly more environmental cost and natural resource depletion than not producing them at all. Compounding the sheer volume of relatively benign alkaline batteries are a host of small button and coin cell batteries that are not alkaline, and due to their size, end up in waste streams headed for the landfill anyway. Button cell batteries, such as those based on silver oxide and zinc, may still contain small amounts of mercury, a heavy metal that was prohibited in the United States in all but button cell batteries by the Mercury-Containing and Rechargeable Battery Management Act of 1996. The bottom line is that almost all of the energy demanded by these small, single-use batteries can, with proper charging strategies and system design, be supplied by rechargeable or wearable energy systems that, in turn, can drastically reduce the negative environmental impacts of wearables and portables.

But, rechargeable batteries are not without their problems either. The dominant rechargeable battery for portables and wearables is based on lithium including lithium ion and lithium polymer technologies. The demand for lithium to produce these batteries is part of the escalating mineral crisis in the world today. While lithium is a metal and not a mineral, extraction from natural reserves requires pumping a rich range of minerals from the earth's crust to generate enough lithium carbonate for extraction. This process uses large amounts of water and often does so in dry, desert-like areas. The resulting evaporative extraction pools required to process lithium can seriously diminish local water supplies, stimulate regional conflicts, and create substantial risk to the safety of local waterways through potential leaks (Katwala 2018). Complicating matters, worldwide production capacity and supplies of cost-effective, accessible lithium are unlikely to keep pace with lithium demand. By 2025, battery demand is expected to dominate lithium demand, surpassing the glass and ceramics industry which is the current front-runner. While there is no fundamental shortage of lithium on planet Earth, there is and will remain a shortage of cost-effective and environmentally acceptable means to mine it (Clean Energy Trust 2018).

Despite the toll that global hunger for batteries takes on the environment, in the short term, wearable solar cell systems do nothing to reduce the number of batteries (and also reduce the corresponding amount of natural resources) used to power an individual's full plethora of portables and wearables. In initial implementations, these systems would simply supply power to existing rechargeable batteries, thus replacing AC charging and the fossil fuel consumption that typically produces AC power. In the long term, however, these systems have the potential to change the way portables and wearables are designed and supported—by offering a nearly continuous source of power available anywhere there is light. The amount of power available at any given point in time can vary tremendously based on the intensity and nature of ambient lighting. Nevertheless, such potential to harvest energy on the go can change the entire paradigm of how portables and wearables are powered. The use of a battery, whether rechargeable or otherwise, presumes no other reliable power source, thus mandating the consistent, reliable, and predictable long-term energy storage that only high-performance batteries can provide. Wearable solar cell systems can reduce the requirements for energy storage and open the door to other energy storage options. For example, supercapacitors can take and deliver charge much faster than batteries and have many more recharging cycles (i.e., longer lifetimes) than regular batteries. Their increased self-discharge times (and lower energy storage periods) compared to traditional batteries can become largely irrelevant in a system where ambient light or other energy is harvested not only to supply power to electronic devices but to compensate for this self-discharge. And, supercapacitors can be made with materials that, unlike many materials used in rechargeable batteries, are both environmentally friendly and relatively abundant (Dyatkin et al. 2013).

1.2.4 Other Impacts

The potential for portables and wearables to improve quality of life is, in part, limited by energy storage, particularly batteries. Wearable solar cell systems have the potential to open up new applications for portables and wearables with significant implications for society. For example, the Tactical Assault Light Operator Suit (TALOS) project initiated by the US military in 2013 endeavored to provide smart, advanced tactical armor to soldiers in close-combat environments. One of the primary goals of TALOS was to reduce injuries and casualties among soldiers while simultaneously improving the effectiveness and efficiency of military operations. In addition to integrating a wide variety of functions including infrared (night) vision and biosensor-based physiological monitoring, TALOS was intended to regulate temperature inside the suit, automatically adjust suit flexibility to provide protection from bullets and other enemy fire, and redirect energy to facilitate more agile movement. The project was recently cancelled because it failed to resolve connectivity and interdependency issues among subsystems and the exosuit simply consumed far too much power given size and weight constraints imposed by conventional batteries (Keller 2019). The development of effective and wearable energy harvesting products, including wearable solar cell systems, is essential to provide greater protection for soldiers, particularly in a world where battle tactics continue to evolve rapidly. Such advanced technologies, while initially developed for the military, are also frequently adapted to support and protect civilian first responders in firefighting, medical emergencies, and natural disaster situations.

Personal healthcare also stands to benefit from greater energy availability among portables and wearables. Power suits that not only sense physiological parameters but also support greater ease of movement have the potential to allow aging individuals to retain their independence much longer and stay out of assisted living for longer, thus reducing the load on the healthcare system and also improving their overall quality of life. Power suits are designed to be worn comfortably and unobtrusively underneath or in place of regular clothing. They contain both sensors and motors to help movement-impaired individuals stand, walk, sit up, and perform other movements that may be otherwise difficult (Williams 2017). Not surprisingly, these types of wearable systems have much greater power requirements than the average fitness tracker or smartwatch and require more creative and supplemental approaches to finding energy to meet those power requirements. Whole body monitoring, which includes tracking muscle and heart activity, logging sitting, standing, and moving postures, and even real-time sensing of environmental exposures among other functions, also requires much more power than most commercially available portables and wearables. Products that provide these kinds of functions cannot gain widespread use without concurrent advancement of energy harvesting technologies, including wearable solar cell systems.

Like many other technological advancements, the full realm of possibilities for portables and wearables has not yet been fully explored. Increasing the availability of energy to power these systems is a critical piece of exploring the full range of positive impacts they can have in a wide range of markets that include not only the military but also extend well beyond it to civilian market sectors.

1.3 Feasibility of Wearable Solar Cell Systems

Can wearable solar cell systems generate enough power to make a difference?

Outdoors, the answer is a resounding yes. The energy that sunlight provides can vary from about 10 W/m² on an overcast day to over 1,000 W/m² on a sunny day (Littlefair 1985). If only half the body's 1.7 m² surface area (Shiel 2018) was covered by a wearable solar cell system and at a dismal efficiency of 10%, between 8.5 and 850 Wh of energy would be generated over a typical 10-hour day. At a minimum, this is sufficient power to support most smartphones, supply power to a laptop computer for several hours, or power many display-free wearables such as hearing aids for multiple days.

Indoors, the problem is a little more challenging. The energy that different artificial lights in classrooms and offices can supply to a wearable solar cell system is generally much lower than what is available outdoors, varying between about 2 and 36 W/m² (National Optical Astronomy Observatory n.d.; Ledke Technology Ltd. n.d.). Over 10 hours a day with a 10% efficient solar system, this amounts to between 1.7 and 30.6 Wh, increasing to 2.7 and 49 Wh for a 16-hour exposure to account for the fact that artificial lighting is not limited by the availability of natural daylight. Even at the lower end of energy production, such a wearable solar cell system produces enough power to adequately support heavy usage of most wearables and light usage of many portables (e.g., smartphones, tablets, laptop computers).

An efficiency of 10% is a conservative estimate of the efficiency by which light energy can be converted to electrical energy. Best-case performance of modern-day solar cells exceeds this conservative estimate by quite a bit (Figure 1.1), especially in more complex solar cell structures such as tandem cells that consist of multiple bilayer cells operating concurrently (i.e., in tandem). While wearable solar cell systems are feasible today even under the relatively low-light energies associated with artificially lit environments, advances in solar cell technology will continue to support more efficient solar cells and systems well into the future.

Portable solar systems have already been introduced into commercial markets, providing between 20 and 120 W of power under ideal sunlight

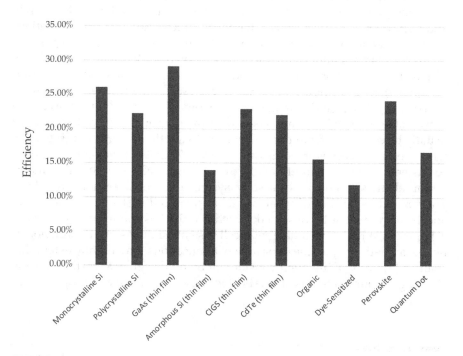

FIGURE 1.1

Energy conversion efficiencies of solar cells. Best research cell efficiencies are shown for bilayer junction solar (PV) cells fabricated in various technologies, based on data from the National Renewable Energy Laboratory (NREL 2019). Tandem cells which consist of multiple bilayer junction cells operating concurrently are not shown but demonstrate efficiencies as high as 39.2%.

conditions. Translated to energy, a 100 W solar panel will produce about 550 Wh/day in a sunny outdoor location in Arizona and about 350 Wh/day in a cloudier outdoor location in western Washington state (Solaris n.d.). Energy production when these panels are exposed to indoor lighting will be significantly less.

1.4 Summary

While the power of sunshine is more than sufficient to energize most portables and wearables using wearable solar cell systems, most individuals spend much of their time indoors away from this rich and abundant source of energy. All is not lost, however, as the power of light is also present indoors in any ambient light, including that generated from fluorescent, incandescent, and light-emitting diode (LED) sources. Understanding the power of light and the capacity of PV solar cells to convert this light to electrical energy begins with an overview of the fundamental behavior of light and basic operation of PV cells (Chapter 2). Efficiencies and other variations in the conversion of light energy to electrical energy are highly dependent on the PV material underlying the construction of a PV cell including silicon (Chapter 3), other inorganic semiconductors (Chapter 4), and other materials (Chapter 5). Since a single PV cell will never a solar panel make, some discussion of array design and management (Chapter 6) is warranted in the context of wearable solar cell systems. Thereafter, the output of a PV cell array must usually be converted from one DC voltage to another before charging a battery or other energy storage device (Chapter 7) that subsequently provides consistent power to a variety of electronic devices (Chapter 8). And, finally, taking all these elements of a solar system into consideration provides a greater understanding of what wearable solar cell systems are capable of and how they must be designed to achieve the best performance possible (Chapter 9). Whether wearable solar cell systems simply serve to calm nomophobic anxiety or open the door to new applications that directly extend life span and quality of life, their consideration and design is a worthwhile and challenging pursuit that is off the beaten track of stationary rooftop designs.

References

Barbaschow, Asha. 2019. "Smartphone Market 'a Mess' but Annual Tablet Sales Are Also Down." ZDNet. February 1, 2019. https://www.zdnet.com/article/smartphone-market-a-mess-but-annual-tablet-sales-are-also-down/.

Best Places. n.d. "Climate in Seattle, Washington." Accessed March 8, 2019. https://www.bestplaces.net/climate/city/washington/seattle.

Boxwell, Michael. 2019. "Solar Irradiance Calculator." In *Solar Electricity Handbook*. London, England: Greenstream Publishing. http://www.solarelectricityhandbook.com/solar-irradiance.html.

Bradford, Abi, and Bret Fanshaw. 2018. "Shining Cities 2018: How Smart Local Policies Are Expanding Solar Power in America." https://patch.com/us/across-america/here-are-americas-winners-solar-energy.

Center for Sustainable Systems. n.d. "Residential Buildings Factsheet." Accessed March 8, 2019. http://css.umich.edu/factsheets/residential-buildings-factsheet.

Clarridge, Christine. 2016. "What You Need to Know about the Winter Solstice: The Days Will Start to Get Longer!" *The Seattle Times*, December 20, 2016. https://www.seattletimes.com/life/what-you-need-to-know-about-winter-solstice/.

Clean Energy Trust. 2018. "Is There Enough Lithium to Feed the Current Battery Market Demand?" *Clean Energy Trust* (blog). February 13, 2018. http://cleanenergytrust.org/enough-lithium-feed-current-battery-market-demand/.

Crees, Alex. n.d. "Energy Rankings: Which States Use the Most Electricity per Household?" Accessed July 15, 2019. https://www.chooseenergy.com/news/article/the-states-that-use-the-most-and-least-amount-of-energy-per-household/.

Dyatkin, Boris, Volker Presser, Min Heon, Maria R. Lukatskaya, Majid Beidaghi, and Yury Gogotsi. 2013. "Development of a Green Supercapacitor Composed Entirely of Environmentally Friendly Materials." *ChemSusChem* 6 (12): 2269–2280. https://doi.org/10.1002/cssc.201300852.

Evans, Simon. 2017. "Solar, Wind and Nuclear Have 'Amazingly Low' Carbon Footprints, Study Finds." Carbon Brief. December 8, 2017. https://www.carbonbrief.org/solar-wind-nuclear-amazingly-low-carbon-footprints.

Holst, Arne. 2018. "Global Notebook Shipments 2016–2018." Statista. August 27, 2018. https://www.statista.com/statistics/818424/global-notebook-computer-shipments/.

International Data Corporation (IDC). 2019. "IDC Reports Strong Growth in the Worldwide Wearables Market, Led by Holiday Shipments of Smartwatches, Wrist Bands, and Ear-Worn Devices." IDC: The Premier Global Market Intelligence Company. March 5, 2019. https://www.idc.com/getdoc.jsp?containerId=prUS44901819.

Katwala, Amit. 2018. "The Spiralling Environmental Cost of Our Lithium Battery Addiction." *Wired UK*, August 5, 2018. https://www.wired.co.uk/article/lithium-batteries-environment-impact.

Keller, Jared. 2019. "SOCOM's Iron Man Suit Is Officially Dead." Task & Purpose. https://taskandpurpose.com/talos-iron-man-suit-dead.

Ledke Technology Ltd. n.d. "What Is Luminous Efficacy? Definition | Flexible LED Display, Curved LED Screen, Soft, Front Service Screen." Accessed July 9, 2019. http://www.ledke.com/what-is-luminous-efficacy-definition/.

Littlefair, Paul J. 1985. "The Luminous Efficacy of Daylight: A Review." *Lighting Research & Technology* 17 (4): 162–182. https://doi.org/10.1177/14771535850170040401.

Markets and Markets. 2014. "Portable Battery Pack Market Worth $10.94 Billion by 2020." https://www.marketsandmarkets.com/PressReleases/portable-battery-pack.asp.

McSweeney, Robert, Rosamund Pearce, and Tom Prater. 2018. "Interactive: The Impacts of Climate Change at 1.5C, 2C and Beyond." October 4, 2018. https://interactive. carbonbrief.org/impacts-climate-change-one-point-five-degrees-two-degrees/.

National Optical Astronomy Observatory (NOAO). n.d. "Recommended Light Levels (Illuminance) for Outdoor and Indoor Venues." https://www. noao.edu/education/QLTkit/ACTIVITY_Documents/Safety/LightLevels_ outdoor+indoor.pdf.

National Renewable Energy Laboratory (NREL). 2019. "Best Research-Cell Efficiency." https://www.nrel.gov/pv/cell-efficiency.html.

NREL. n.d. "PV Watts Calculator." Accessed July 15, 2019. https://pvwatts.nrel.gov/ pvwatts.php.

Power Bank Guide. 2019. "Best Smallest and Lightest Power Banks for Handbags & Travel." *PowerBankGuide* (blog). July 7, 2019. https://www.powerbankguide. com/best-ultra-compact-powerbanks/.

Shiel, William C. 2018. "Definition of Body Surface Area." MedicineNet. https:// www.medicinenet.com/script/main/art.asp?articlekey=39851.

SocialCompare. 2019. "Apple IPhone Product Line Comparison." June 17, 2019. http:// socialcompare.com/en/comparison/apple-iphone-product-line-comparison.

Solaris. n.d. "Solar Panel Ratings Explained." Solaris. Accessed July 13, 2019. https:// www.solaris-shop.com/blog/solar-panel-ratings-explained/.

The World Bank. 2014. "Electric Power Consumption (KWh per Capita) | Data." https://data.worldbank.org/indicator/EG.USE.ELEC.KH.PC?locations=KE.

Veloxity. 2017. "Cell Phone Battery Statistics 2015 2016 2017." https://veloxity. us/2015-phone-battery-statistics/.

Walden, Stephanie. 2018. "The 'Indoor Generation' and the Health Risks of Spending More Time Inside." *USA Today*. May 15, 2018. https://www.usatoday.com/ story/sponsor-story/velux/2018/05/15/indoor-generation-and-health-risks- spending-more-time-inside/610289002/.

Williams, Brett. 2017. "These Power Suits Could Make Your Grandma's Life Much Better." Mashable. January 11, 2017. https://mashable.com/2017/01/11/ superflex-powered-clothing/.

Xará, Susana, Manuel Fonseca Almeida, and Carlos Costa. 2015. "Life Cycle Assessment of Three Different Management Options for Spent Alkaline Batteries." *Waste Management* 43 (September): 460–484. https://doi.org/10.1016/j. wasman.2015.06.006.

Yildirim, Caglar, and Ana-Paula Correia. 2015. "Exploring the Dimensions of Nomophobia: Development and Validation of a Self-Reported Questionnaire." *Computers in Human Behavior* 49 (August): 130–137. https://doi.org/10.1016/j. chb.2015.02.059.

2

Fundamentals

Although there is more than one approach to harness solar or light energy, the photovoltaic (PV) approach is by far the most prevalent. PV technologies rely on the PV effect to convert solar or light energy to electricity. The PV effect refers to the excitation of electrons within a material by incoming light (irradiation) and is closely related to the photoelectric effect, in that both involve the excitation of electrons from low-energy to high-energy states. In the photoelectric effect, however, an electron is excited to the vacuum level or free space (i.e., fully freed from the material), whereas in the PV effect, the energized or excited electron remains linked or associated with the material where it was originally harbored in its resting or low-energy state.

A wide range of designs and strategies are used to exploit the PV effect to produce usable electrical energy. But, regardless of their design, structure, or component materials, PV cells behave in very similar ways. Light travels into (irradiates) the PV cell. Light is converted to energized electrons. These high-energy electrons are then collected at the edges (contacts) of the PV cell and transferred to an external circuit for conversion to useful electrical energy.

How well these processes take place inside a PV cell is dependent on the properties of incoming light, the opportunities for light to energize electrons, and the capacity of the PV cell to retain and collect energized electrons rather than losing them to recombination. These fundaments of PV cell behavior are discussed in this chapter in the context of the most successful commercially available PV material to date—monocrystalline silicon.

2.1 Light

PV cells require light to function and be useful. Light is described by both basic properties and by behavioral characteristics that can be measured either in absolute (i.e., independent of the human eye) or relative (i.e., dependent on the sensitivity of the human eye) terms.

Relative or photometric measurements of light are done in terms of units such as lumens, lux, and foot-candles that scale light based on how the human eye responds to that light. For example, 1 watt (W) of light at 555 nanometers (nm) (i.e., green light) is equivalent to 683 lumens, while at 510 nm, the same amount of light (1 W) is approximately 340 lumens because the human eye is only about half as sensitive to 510 nm light as 555 nm light. Since the sensitivity of the human eye to light is largely irrelevant to the performance of PV cells, these photometric units are equally irrelevant to PV cell operation and performance. Instead, a discussion of PV cells is better suited to radiometric or absolute units such as watts to describe input light energy. Unlike lumens, 1 W of light at 500 nm is the same as 1 W of light at 555 nm—1 J/sec of energy.

2.1.1 Basic Properties of Light

Light can be described by basic properties of intensity, frequency, wavelength, and velocity.

Intensity simply refers to the amount of light energy. The most common units used in the description of intensity in the context of PV cell behavior are the *radiant energy* (energy of the light in Joules or J), the *radiant flux* or power (energy of the light per unit time in watts), and the irradiant flux density or *irradiance* (radiant flux reaching a surface per unit area in W/m²). The term intensity is often used more informally to describe multiple other properties of light which can be confusing.

The color of a light source (including the sun) is associated with the wavelength or wavelengths of light it emits. When depicted as a wave (Figure 2.1), light has a wavelength (λ) represented by the distance between two adjacent peaks on the wave and expressed in units of nm or micrometers (μm).

Wave speed is the number of peaks that pass by a fixed point in space per time (in m/sec), and the velocity (v) expresses both the speed and direction

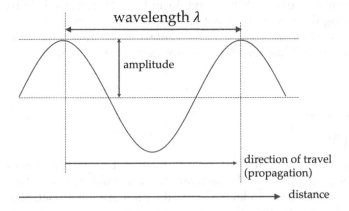

FIGURE 2.1
Basic properties of light. The amplitude of a wave is the distance between a resting position and the maximum displacement of the wave from that resting position in a direction perpendicular to the direction of wave travel. The wavelength (λ) is the distance between neighboring identical points on the wave such as the distance between adjacent peaks. Frequency (f) is the number of waves that move past a specific point in space every second and wave speed or velocity (v) is simply the product of wavelength and frequency in the direction that the wave is travelling.

of the wave. The frequency of the wave (f) is related to the wavelength and velocity (v) as follows:

$$f = v / \lambda \tag{2.1}$$

In air, the velocity of a light wave is approximately equal to the velocity of light in a vacuum (c) and has a value of 3×10^8 m/sec. In other denser media, the velocity of light is related to the refractive index (RI) of that material and the speed of light in a vacuum:

$$v = c / RI \tag{2.2}$$

The speed and velocity of light in dense media are always slower than the speed and velocity of light in air or in a vacuum, but the frequency of light remains constant regardless of the medium through which the light is travelling. Most light sources are a combination of light travelling at different frequencies and having different wavelengths. The laser comes closest to being composed of a single wavelength of light with a very narrow bandwidth, while sunlight is broad, with wavelengths ranging from ultraviolet through the visible region into the infrared spectrum. And, other light sources, such as LEDs, contain wavelengths with a broader bandwidth than lasers but not nearly as broad as sunlight (Figure 2.2).

Regardless of the breadth of the spectrum emitted by a light source, the component wavelengths add together to produce the color observed by the human eye. Different combinations of wavelengths can add up to the same perceived color. Therefore, perceived color alone does not uniquely identify the wavelengths of light that are present in a light source. Light sources that contain a wide range of wavelengths in the visible region are typically perceived to be white, but those that contain a narrower range of wavelengths can look like any of the colors in the rainbow. For example, green light-emitting diodes (LEDs) are dominated by wavelengths between 550 and 600 nm, and their color is very clearly observed as green. In contrast, sunlight contains a combination of visible light with wavelengths between 400 and 750 nm that when combined, make sunlight look white or near white. Sunlight also contains wavelengths outside the visible light region including infrared light up to 2500 nm and ultraviolet light (wavelengths shorter than 400 nm).

2.1.2 Behavioral Characteristics of Light

As light travels through different materials or media and encounters obstacles, it can exhibit a wide range of behaviors. The most relevant behaviors to a discussion of PV cells are reflection, transmission, and absorption. Light that bounces back from a surface without being transmitted or absorbed is reflected. Some light is reflected in a particular direction, while other light (particularly on rough surfaces) bounces back in a variety of directions and is said to be scattered.

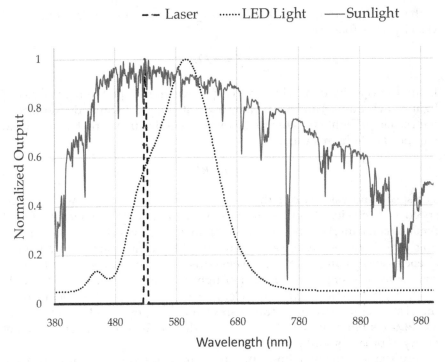

FIGURE 2.2
The color of light. The color or spectral composition of several common light sources including terrestrial sunlight based on AM1.5 (National Renewable Energy Laboratory [NREL] n.d.).

Regardless of whether light is scattered or reflected from the surface of a PV cell, all such light is lost as a potential source of energy to the cell. Of the light that is not reflected, some is transmitted unchanged through the PV cell and some is absorbed by the PV cell. For silicon and other semiconductors, absorbed light is converted to pairs of free current carriers called electron-hole pairs.

Light that is successfully absorbed by a material is only absorbed in discrete energy packets called photons. The energy of a single photon E is given by the Planck-Einstein relationship:

$$E = hc / \lambda \tag{2.3}$$

where h is Planck's constant (6.63×10^{-34} J-sec or 4.14×10^{-15} eV-sec), c is the speed of light (3×10^8 m/sec), and λ is the wavelength in meters. A material must absorb an integer number of photons. If energy equivalent to 3.5 photons of light is incident on a material, the maximum number of photons that can be absorbed is 3; the remaining 0.5 photon worth of energy remains unabsorbed and is transmitted through the material.

The irradiance of input light determines the maximum possible power that can be produced by a PV cell. In an ideal situation, a 1 m² PV cell exposed to

1 W/m² of input light would produce 1 W of output power. In reality, however, there are many different losses that occur in a real PV cell that significantly diminish how much input light can be converted to output power. For instance, a typical basic PV cell made of (monocrystalline) silicon produces a maximum of about 0.25 W of output power for every watt of input sunlight. Where does all the wasted power go? By far, the greatest "waste" of power incurred by PV technology occurs because of the mismatch between how and how much the PV cell absorbs light and how and how much light source emits light as a function of wavelength.

2.2 PV Materials

Consider the behavior of monocrystalline silicon operating as a PV cell to harness light or solar energy. At a temperature of absolute zero (0 K) and zero illumination (no light) conditions, all the electrons in the semiconductor are at low energy levels in the valence band of the silicon (Figure 2.3a). As the temperature increases, some electrons gain thermal energy and move into the higher energy conduction band (Figure 2.3b) but not into the bandgap. The bandgap is a fundamental characteristic of the semiconductor and represents a range of energies forbidden to electrons. In reality, some energy states do exist in the bandgap, but for purposes of this discussion, the bandgap energies will be assumed to be unoccupied.

When silicon is exposed to light, a photon can be of a wavelength whose energy is smaller than the bandgap energy (Figure 2.3c). In this case, an

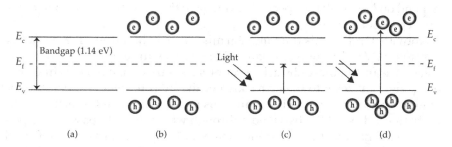

FIGURE 2.3

Energy bands in pure (undoped) silicon. (a) At absolute zero (0 K), no current carriers (i.e., free electrons or holes) are present; (b) at higher temperatures, some free current carriers are available; (c) in the presence of incoming light whose energy is lower than the band gap, the number of free current carriers does not change; and (d) in the presence of incoming light whose energy is greater than the bandgap energy, the number of free current carriers increases. The Fermi level (E_f) indicates the ratio of free electrons to free holes in a semiconductor. In the absence of light, pure intrinsic silicon contains equal numbers of free electrons to free holes, which puts the Fermi level in the middle of the bandgap.

electron is unable to cross the bandgap. It remains in its resting energy state and the photon is not absorbed. Using the Planck-Einstein relationship (Equation 2.3), the maximum wavelength of light which can be absorbed by silicon (bandgap of 1.14 eV) is 1.09×10^{-6} meters or 1,090 nm. Light longer than 1,090 nm does not have sufficient photon energy to cross the bandgap and is not absorbed. This means that most infrared light irradiating a silicon PV cell cannot be absorbed. Sunlight has a significant infrared light component which limits silicon's ability to absorb and convert sunlight into electrical energy (Figure 2.2).

Alternatively, light can be absorbed and provide just enough energy (per photon) for an electron to cross from the upper edge of the valence band to the lower edge of the conduction band. This is the most efficient absorption of light and results in no wasted energy, but it only occurs at a specific wavelength. More likely, light that is absorbed has more than the minimum energy needed for an electron to cross the bandgap (Figure 2.3d). In this scenario, the overexuberant electron loses energy as heat as it drops down to the lower energies in the conduction band. This excess energy, lost as heat, represents an additional source of inefficiency in the semiconductor.

Unfortunately, the semiconductor bandgap which enables light to be harnessed into energy also leads to significant losses at all energies that are greater or less than the bandgap energy. Thus, it is no surprise that for sunlight that contains a broad range of wavelengths from ultraviolet to near-infrared, silicon, at best, can convert a maximum of 32% of incoming light to output power, thus losing a minimum of 68% of incoming sunlight energy to heat or lack of absorption (Rühle 2016). The 32% ultimate efficiency can be thought of as a result of the fundamental mismatch between how sunlight behaves and how a PV cell reacts or responds. Ultimate efficiencies are different for artificial light sources than for sunlight and depend on how well the spectral coverage of the PV cell matches that of the artificial light source.

Notwithstanding this fundamental mismatch, light travelling on its merry way into a PV cell encounters many other obstacles to being converted to electricity. The first of these obstacles occurs at the surface of the cell. While most of the light makes it to the active area of the PV cell, some is reflected. The amount that is reflected rather than transmitted into the cell is related to the refractive indices of the two media through which the light passes. The percentage of light that passes through material 1 (typically air) and is reflected from a PV cell whose surface layer is made of material 2 is approximately:

$$R = 100\% \times \left[\frac{n_2 - n_1}{n_2 + n_1} \right]^2 \tag{2.4}$$

where n_1 is the refractive index of material 1 and n_2 is the refractive index of material 2. For a silicon PV cell with refractive index of about 4 that is exposed to air at a wavelength of 600 nm, approximately 36% of the light is reflected

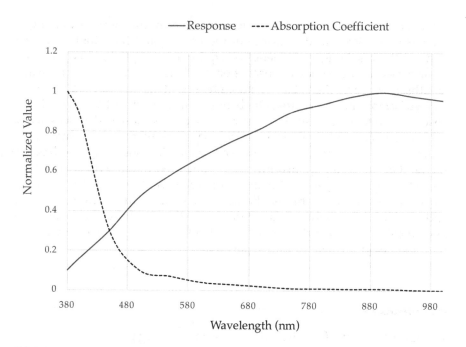

FIGURE 2.4

The response of silicon to light. Shown are the absorption properties of silicon at 300 K based on data from (Green 2008) and the response of silicon to light at various wavelengths based on data from (Boivin et al. 1986). The maximum absorption coefficient of silicon is approximately 106 cm^{-1} at short wavelengths (i.e., those below 400 nm) and indicates how thick the silicon must be to absorb all the light at that wavelength. The response indicates how the photocurrent generated by light of the same irradiance varies with wavelength.

from the surface and is lost to the PV cell. To reduce reflection losses, an antireflective coating can be deposited on top of the surface of the PV cell. In the case of silicon, an antireflective silicon dioxide coating (of refractive index between 1.4 and 1.55) would reduce initial reflection losses to as low as 2.8%.

Compounding these losses, long wavelengths of light may not be fully absorbed in a thin PV cell, because their absorption coefficients are quite small compared to that of shorter wavelengths (Figure 2.4). The amount of light absorbed is dependent on both the wavelength of light travelling through the cell and also the depth the light that travels into the cell. If the active area of a PV cell begins at distance z_1 (cm) and extends to distance z_2 (cm) underneath the surface of the cell, the total amount of light available to be absorbed can be described by application of Beer's law:

$$I_{available} = I_o \exp(-\alpha z_1) - I_o \exp(-\alpha z_2) \tag{2.5}$$

where I_o is the intensity of light incident on the PV cell surface (after reflection losses are taken into account) and α is the absorption coefficient of the PV cell material (in cm^{-1}) and is dependent on wavelength. Shorter wavelengths are absorbed closer to the surface of the PV cell and have higher absorption coefficients. Longer wavelengths require more depth into the PV cell to be absorbed and have lower absorption coefficients. And, very long wavelengths are not absorbed at all because photons of light at those wavelengths lack sufficient energy to cross the bandgap in the active PV material. The absorption behavior of silicon (Figure 2.4) exhibits these differences across wavelength as do other semiconductors as well as alternative PV materials.

In summary, most PV cells experience a tremendous loss of light well before the light even has an opportunity to be converted into electrical current in the active layers of the cell. Some light is reflected. Some light is absorbed by non-active layers of the cell. Some light is transmitted through the cell entirely. Although these losses are minimized in commercial PV cells, they nevertheless have an impact on the overall efficiency of the solar panel, module, or system to which they are a part.

2.3 Conversion of Light into Electrical Energy

PV cells convert light into electrical energy by stimulating electrons to absorb light energy and thereby move to energy levels higher than their resting state. Although the origin and destination of electrons moving from lower to higher energy levels can look different in different PV technologies, the underlying PV behavior is the same.

PV cells based on inorganic semiconductors dominate global PV markets. Over 90% of PV production around the world is based on crystalline silicon semiconductor (Fraunhofer Institute for Solar Energy Systems 2019), while in the United States, more than half of PV production is based on mono- or multicrystalline silicon with another 24% based on the semiconductor cadmium telluride (Platzer 2015). Because of its market dominance and technological maturity, the PV behavior of crystalline silicon is well understood and makes a logical point of departure for understanding other PV cell technologies.

Silicon has an atomic number of 14, is part of group 14 in the periodic table, and is the second most abundant element in the earth's crust at about 28% of the total by mass. The silicon atom has two electrons in its first full shell, eight in the second full shell, and four electrons in the last (partially empty) shell. To achieve a highly desirable and stable full outer shell, silicon will bond with four neighboring silicon atoms for a total of eight electrons in its outer shell (Figure 2.5a). In this state, silicon is stable and relatively insulating, with few electrons available in the conduction band to conduct current. When light of appropriate energy irradiates silicon, each packet (photon) of light energy that is successfully absorbed by the silicon causes an electron to move from the valence band to the conduction band of energies. In so

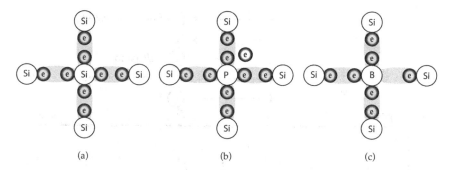

FIGURE 2.5
Silicon. (a) Undoped silicon; (b) extra electron in n-type silicon doped with phosphorous atoms; (c) extra hole in p-type silicon doped with boron atoms.

doing, an energized (excited) electron leaves behind a vacancy (or hole) in the valence band that serves as a mechanism for electrons in the valence band to "hop" from vacancy to vacancy and thereby conduct current. In a single layer of plain silicon, however, the excited electrons in the conduction band do not last long enough (i.e., remain energized) to be collected at the contacts of the PV cell. Instead, most of the electrons are lost and return to the valence band where they recombine with holes.

For any PV cell, whether a single-, double-, or multiple-layer cell, the ability to effectively collect free carriers (electrons and holes) is expressed as the external quantum efficiency (EQE) of the PV cell:

$$EQE = \frac{Electrons\,/\,sec}{Photons\,/\,sec} \tag{2.6}$$

The EQE of single-layer silicon is quite poor because many electron-hole pairs are never collected due to recombination, but it can be increased by forming a junction between two different types of silicon: one p-type and one n-type. The p-type silicon (Figure 2.5c) is created by the addition of atoms (acceptors) that have three electrons in the outer shell (e.g., boron), and create one absence of an electron (i.e., a hole) in the valence band of the crystal per acceptor atom added. The n-type silicon (Figure 2.5b) is created by the addition of atoms (donors) that have five electrons in the outer shell (e.g., phosphorous) and create one additional electron in the conduction band of the crystal per donor atom added.

The addition of dopant atoms (donors or acceptors) increases the conductivity of the silicon. For each donor atom added to undoped silicon, the conductivity is increased by the presence of one additional current-carrying electron. For each acceptor atom added, the conductivity increases by the presence of one additional (but slower and less mobile) current-carrying hole. The conductivity of the resulting semiconductor can be adjusted or tuned by controlling the amount of dopant atoms in the semiconductor.

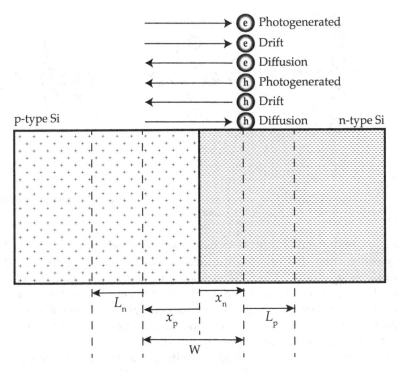

FIGURE 2.6

Electrons and holes in p-n junctions. Diffusion causes electrons to flow from right to left (n-side of the junction to p-side of the junction) and holes from left to right. Drift counteracts diffusion and causes electrons to flow left to right and holes from right to left. The net result is a dynamic equilibrium where diffusion and drift balance one another at the built-in potential. Incoming light generates electron-hole pairs throughout the p-n junction. Electron-hole pairs within the depletion region and within one diffusion length of its boundaries are far less likely to recombine and more likely to be successfully collected for the production of useful electricity.

When p-type and n-type materials are joined together, however, they create a more efficient PV cell via the electric field that appears at the junction between the two materials. Consider a simple p-n junction where n-type silicon is joined to p-type silicon (Figure 2.6). At the moment the two materials are joined, there is no electric field at the junction and both p-type and n-type materials are electrically neutral. However, electrons soon diffuse from the electron-rich n-side of the junction to the electron-poor p-side. In so doing, these electrons take their negative charge with them and leave a positive charge behind. This separation of charges (positive on the n-side and negative on the p-side) stimulates the drift of electrons back to the n-side of the junction where they originated as the opposite charges attract one another. Diffusion and drift continue in a dynamic equilibrium that maintains a level of negative charge on the p-side of the junction and positive charge on the n-side of the junction.

The junction is characterized by two primary parameters. The first is physical area near the p-n junction over which electrons have recombined with holes, leaving no free charge carriers. This physical area is called the depletion region and is dependent on the type of material and the doping on each side of the junction. The total width of this region is represented by the following relationship:

$$W = \sqrt{\frac{2\epsilon\epsilon_o}{q}(\phi_o)\left(\frac{1}{N_a} + \frac{1}{N_d}\right)} \qquad (2.7)$$

where ϵ is the relative permittivity of the semiconductor (e.g., 11.7 for silicon), ϵ_o is the permittivity of free space (8.85×10^{-14} F/cm) ϕ_o is the built-in potential associated with the p-n junction, q is the electronic charge (1.6×10^{-19} coulombs), and N_a and N_d are the doping levels on the p-side and n-side of the junction, respectively, in units of atoms/cm^3. The portion of the depletion region width that lies on the n-side of the junction is:

$$x_n = W\frac{N_a}{N_a + N_d} \qquad (2.8)$$

and the portion of the depletion region width that lies on the p-side of the junction is:

$$x_p = W\frac{N_d}{N_a + N_d} \qquad (2.9)$$

The built-in potential across the p-n junction can be calculated as:

$$\phi_i = \frac{kT}{q}\ln\frac{N_aN_d}{n_i^2} \qquad (2.10)$$

where T is the temperature in kelvin, k is the Boltzmann constant (1.38×10^{-23} J/K) and n_i is the intrinsic carrier concentration (approximately 1×10^{10}/cm^3 for silicon at room temperature). The built-in potential can be visualized by looking at the energy band structure of p-type and n-type silicon before and after a p-n junction is formed. The band structure can be characterized by the conduction band energy (E_c), valence band energy (E_v), the Fermi level (E_f), the vacuum level (E_o), and the work function (ϕ). The vacuum level represents the energy of an electron that is unattached to any material (i.e., in a perfect vacuum) and the work function represents the energy difference between the Fermi and the vacuum level. The Fermi level is the energy level that has a probability of 0.5 of being occupied and reflects the energy that is

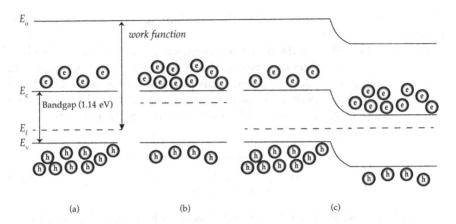

FIGURE 2.7
Energy view of p-n junctions. Energy band structures for (a) p-type silicon, (b) n-type silicon, and (c) a silicon p-n junction. Other semiconductors have different bandgaps (E_g) but exhibit similar behavior when fabricated into p-n junctions.

equally likely to be filled by an electron as a hole. In most semiconductors, the Fermi level is in the bandgap that, by definition, contains no or very few electrons or holes. A more practical definition of the Fermi level for a semiconductor is that it represents the average energy of free electrons and holes in the semiconductor, and its position in the bandgap reflects the ratio of free electrons to free holes. For a pure undoped (intrinsic) semiconductor, E_f sits in the middle of the bandgap (Figure 2.3). Because p-type semiconductors are dominated by free (current-carrying) holes, E_f sits well below the middle of the bandgap and close to the valence band energy in a p-type semiconductor (Figure 2.7a), and in n-type semiconductors, the opposite is true (Figure 2.7b).

When the p-type and n-type silicon are joined together, however, the Fermi levels must be continuous across the junction, causing the energy levels (E_c and E_v) on one side of the junction to bend into E_c and E_v on the other side of the junction (Figure 2.7c). This band bending creates an energy hill whose height is equal to the built-in potential (ϕ_{bi}) of the p-n junction and enables electrons to flow down the energy hills and holes to float up the hills. Under conditions that preserve the energy hills shown in the p-n junction in Figure 2.7c, the holes would float upward to the left through the p-side of the junction and the electrons would flow downward to the right through the n-side of the junction.

The built-in potential produces a maximum electric field (E) across the junction that is quite large.

While the built-in potential varies between about 0.56 and 1.04 V from lightly doped to heavily doped silicon, the depletion region width is quite small (on the order of tenths of micrometers), which results in electric fields on the order of tens of thousands of volts per cm.

Light irradiates the entire p-n junction. When it is absorbed, a single photon creates both an energized electron in the conduction band and a hole remaining in the valence band. Together, the free electron and free hole are called an electron-hole pair. Far away from the p-n junction, the silicon acts just like a single-layer PV cell. In the absence of any significant electric field, electrons and holes quickly recombine and are unlikely to be collected at the contacts on either end of the PV cell. However, closer to the junction (i.e., inside the depletion region or within one diffusion length of it) where the electric field is quite large, electrons and holes generated by light are pushed away from each other by the built-in electric field and can more readily travel to their respective contacts. Electrons move to the contact on the n-side of the junction and holes move to the contact on the p-side of the junction (Figure 2.6).

Silicon and other semiconductor PV cells capitalize on the built-in electric field and depletion region that comes about when oppositely doped semiconductors are joined together. But, in one way or another, all PV cells rely on this same basic behavior. Once electron-hole pairs are generated, a force of some kind must be present in the PV cell to effectively separate electrons from holes and then transport them to their respective contacts while losing as few as possible to recombination. Silicon p-n junctions do this quite well with internal quantum efficiencies approaching 100% (i.e., ratio of electrons generated per second to absorbed photons per second) at visible light wavelengths (Holman et al. 2013) and overall energy conversion efficiencies (i.e., ratio of power generated to incoming light power) reaching as high as 26.1% (NREL 2019). Silicon has a maximum possible energy conversion efficiency for simple two-layer p-n junction PV cells of 32% (Rühle 2016).

2.4 Advanced PV Designs

In general, single-layer PV materials are inefficient because most electrons and holes recombine before they can be collected and converted to electrical energy. In a bilayer PV cell (e.g., the semiconductor p-n junction), recombination is reduced, and many more electron-hole pairs generated by light are successfully collected. While the design of bilayer PV cells can be optimized to obtain high energy conversion efficiency, other architectures beyond the bilayer design are also possible for more efficiently harnessing light. Two of the more popular approaches to advancing past the capabilities of the bilayer design are the tandem (multiple-junction) PV cell and the use of concentrators.

2.4.1 Tandem Cells

Multiple p-n junctions can be stacked together and operated in conjunction with one another to collect and harness more incoming light. In this tandem cell structure, bilayer junctions are stacked, one on top of the other, with the lower bilayer junction (i.e., the junction further away from the incoming light) collecting low-energy photons that, by their very nature, are absorbed deeper into the cell and the upper bilayer junction collecting high-energy photons that are more fully absorbed closer to the surface of the PV cell. These tandem cells can be electrically connected in series, but a series connection requires that currents through the cells be equal. A mismatch in cells can cause the current in one to decrease to match the other, thus resulting in one or more cells operating outside of the optimal (maximum) power point. Alternatively, tandem cells can be operated individually without being connected in series. While requiring more wires than series-connected cells, individual tandem cells are more flexible both in their design and in the light sources they can accommodate at optimal absorption and efficiency. PV cells using two-cell tandem designs have maximum possible (ultimate) efficiencies of 47% (Bremner, Levy, and Honsberg 2008) and research cells using such tandem designs have demonstrated efficiencies as high as 32.8%, 37.9%, and 39.2% for two-, three-, and four-cell designs, respectively (NREL 2019).

2.4.2 Solar Concentrators

Concentrating light into a solar cell also increases PV cell efficiency. Fresnel lenses and other optics can be used to gather light from an area that is greater than the surface area of a PV cell. The optics then focus or concentrate the collected light directly onto a PV cell, thus increasing the overall irradiance of the cell. In a traditional solar panel installation that involves many, many PV cells and solar panels connected together, concentrator optics offer a less expensive alternative since concentrating optics often cost less per unit area than PV cells. In a concentrator system, fewer PV cells and solar panels can be used to collect essentially the same amount of light. However, solar concentrators require direct light to operate efficiently. Direct light refers to light that is received on a straight line between the sun and the collecting device (i.e., concentrating optics). Thus, concentrator systems require tracking devices to adjust the orientation of the concentrating PV cells throughout the day in order to maximize exposure to direct sunlight (Philipps et al. 2015). The need for direct light limits concentrator systems to sunshine-rich regions of the world, but these concentrator systems have demonstrated research cell efficiencies as high as 46% (NREL 2019). However, being limited to direct light also makes concentrator approaches impractical for wearable PV systems. A wearable PV is unlikely to receive much direct light and trackers installed on such small systems are impractical and prohibitive in terms of both cost and weight. Frequent changes in orientation inherent to mobile and wearable systems also complicate efficient tracking and concentrating of available light.

2.4.3 Other Strategies

Strategies to reduce recombination are also used to improve PV cell efficiency. While bulk p-n junctions work well for silicon and other semiconductors because of their long diffusion lengths, the diffusion lengths for some PV materials are very short, thus limiting the active area for collecting free electrons and holes. In such materials (e.g., organic PV cells), dispersed junctions work better. In a dispersed junction, an electron acceptor and electron donor material are blended together to form a single layer of many, many short junctions. With such short junctions, the distance that a photogenerated electron must travel to reach a junction where it can be separated from its corresponding hole is drastically reduced. As a result, fewer electrons and holes recombine and the PV cell is able to perform at reasonable efficiencies (Nelson 2003).

In still other PV cell technologies, the use of a mesoporous layer is advantageous to collecting light and converting it efficiently to electrical energy. Mesoporous layers contain pores that are typically between 2 and 50 nm in diameter. These pore sizes allow molecules within the layer to be coated with photosensitive dyes or other materials while still allowing ions to travel through the layer to replenish electrons that have been transported out of the PV cell for collection and conversion to usable current (McEvoy 2003).

2.5 Performance of PV Cells

Multiple characteristics speak to the superiority of one PV cell over another and much of such an evaluation is dependent on the application. Far and away, however, the most popular metric used to compare one PV technology to another at a single-cell level is the energy conversion efficiency, given by:

$$\eta = \frac{P_{out}}{P_{in}} \tag{2.11}$$

where P_{out} is the electrical power generated by the PV cell and P_{in} is the incoming optical power, both measured in units of watts. The energy conversion efficiency η is unitless and is often simply called efficiency.

The output power of a PV cell is not constant, but instead varies with the current through and the voltage across the cell, both of which vary with the load attached to the cell. The PV cell voltage can vary between 0 and the open circuit (zero current) voltage (V_{oc}) and the current can vary between 0 and the short circuit (zero voltage) current (I_{sc}). The power produced by the PV cell under open circuit conditions is:

$$P_{out} = I_{cell}V_{cell} = 0 \times V_{oc} = 0 \tag{2.12}$$

which is not useful for the production of electricity. Similarly, the PV cell is not useful operating at short circuit conditions where the output power is also zero, because the cell voltage is zero. Somewhere in between these two extremes is the point of maximum power production or MPP (Figure 2.8). Ideally, a PV cell should operate at its MPP at all times to ensure maximum power production under any illumination and environmental conditions.

In practice, finding the MPP requires varying a load resistance attached to the cell from 0 ohms to very large values of resistance while the cell is irradiated with light. At a load of 0 ohms, the cell operates under short circuit conditions and the current flowing through the cell is I_{sc}. At a very large load resistance, the cell operates under open circuit conditions and the voltage across the cell is V_{oc}. Somewhere in between these two extremes lies the MPP that is used to assess the (maximum) energy conversion efficiency of the PV cell. Research cell efficiencies provided by NREL (NREL 2019) as cited throughout this book are energy conversion efficiencies based on sunlight (AM1.5) as the input light and on a P_{out} corresponding to the MPP. Energy conversion efficiencies for light sources other than terrestrial sunlight will be different.

Energy conversion efficiencies are fundamentally limited to values well below 100%. For a single bilayer PV cell, the maximum achievable efficiency is the ultimate efficiency (also known as the Shockley-Queisser limit) of 33.7% (Rühle 2016). The ultimate efficiency is only achievable when the bandgap of the underlying PV materials is optimally matched to the incoming light source. For sunlight, this optimal match occurs at a bandgap of around 1.34 eV (Rühle 2016).

In addition to the energy conversion efficiency, another parameter relevant to evaluating PV cell performance is the fill factor (FF). The FF can be calculated as:

$$FF = \frac{P_{out,max}}{V_{oc}I_{sc}} \tag{2.13}$$

where $P_{out,max}$ is the power produced by the cell at the maximum power point (Figure 2.8) and V_{oc} and I_{sc} are the open circuit voltage (in volts) and short circuit current (in amperes) for the cell, respectively. The FF provides an indication of the quality of the PV cell and ranges from 50% to over 80% in modern PV cells. Monocrystalline silicon PV cells have demonstrated FFs as high as 83% (Polman et al. 2016).

To optimize the performance of a PV cell, the bandgap of a PV material should be matched to the input light source, reflection losses should be minimized, and the load to the PV cell should be adjusted so that cell operates at MPP no matter what the input light conditions are. A PV cell operating at its best, whether as a single device or as part of a larger system containing many PV cells, is operating at its MPP all the time. In practical operation, however, no PV cell operates at its best all the time.

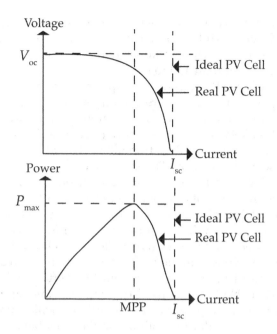

FIGURE 2.8
Power production in a PV cell. As current through a PV cell decreases, voltage increases until no current flows and the voltage reaches a maximum at the open circuit voltage V_{oc}. At zero voltage, the current reaches its maximum at the short circuit current I_{sc}. The maximum product of voltage and current is the maximum power point (P_{max}).

2.6 Shading and Other Irregularities

A wide range of environmental, manufacturing, and design variables can cause a PV cell to operate at less than optimal conditions. Very rarely do PV cells operate in isolation as single-cell systems. But, the characteristic curves of PV cells, even when designed and manufactured to be identical, are different from batch to batch and cell to cell. Manufacturing variation means that under identical input irradiation and output load, only some PV cells will be operating at their MPP. The only way to ensure that each cell in an array is operating at its best is to integrate MPP control into every cell, but the space, power, and cost involved in doing so may be prohibitive. Further complicating matters, in many PV cell arrays, cells are connected in series, which requires that all cells conduct the same current. Thus, even under identical irradiance conditions, these series-connected cells (strings) will not be able to operate at their individual MPPs and the collective MPP of the string will be less than the sum of MPPs in each of the individual cells. Environmental irregularities in some cells and not others have similar effects on power production as manufacturing variations do. Shading, soiling, and aging are three common sources of environmental irregularity that can cause PV cells to operate at different MPPs. These environmental irregularities can be so extreme as to create hot spots in an array of PV cells that can steal power from the system rather than add to it.

The performance of a PV cell is also compounded by heating. Conventional rooftop solar panels heat up in direct sunlight which decreases their open circuit voltage and as a result, their maximum power output. Self-heating effects mean that solar panels frequently operate well above the ambient temperature with efficiencies that drop off on the order of 0.26% per degree Celsius increase in temperature (Panasonic n.d.).

In stationary PV installations, all of these factors (heat, soiling, shading, aging, manufacturing variation) come into play and affect the overall power production of a solar cell system. In wearable PV systems, some of these potential sources of degraded performance are minimized. Wearable systems are less susceptible to the negative effects of self-heating because they are less frequently exposed to the high ambient temperatures and relatively high levels of direct sunlight experienced by rooftop and other outdoor installations. Wearable PV systems also spend less time outdoors and have less frequent exposure to poor weather conditions, which also reduces performance losses that result from soiling. On the other hand, many wearable PV systems are more vulnerable to more frequent and more variable degrees of shading because they are mobile rather than stationary. Furthermore, wearable PV systems are often made to be flexible in order to conform to a wide range of irregular topologies and are therefore more prone to irregularities generated over time from aging, cracking, and other damage.

2.7 Summary

Regardless of their component materials or architecture, PV cells rely on the PV effect to convert incoming light into free current carriers that can be collected and converted to useful electricity. The most mature PV technologies use monocrystalline silicon fabricated in a two -layer structure called a p-n junction. Single silicon p-n junctions have been demonstrated with energy conversion efficiencies over 26% and can potentially reach efficiencies of about 32%. Multiple two-layer structures or silicon p-n junctions can surpass 32% efficiency and are stacked on top of one another in what are called tandem cells.

Despite the dominance of crystalline silicon in commercial solar cells, other forms of silicon and other semiconductors have been developed in a second generation of PV cells based on thin films. Entirely different materials and alternative structures make up another, third generation of PV cells that like second-generation cells, may eventually compete with silicon in stationary, mobile, or wearable PV applications. These first-, second-, and third-generation approaches to converting light energy to electrical energy are discussed in more detail in Chapters 3, 4, and 5, respectively.

References

Boivin, L. Philippe, Wolfgang Budde, C. X. Dodd, and S. R. Das. 1986. "Spectral Response Measurement Apparatus for Large Area Solar Cells." *Applied Optics* 25 (16): 2715–2719. https://doi.org/10.1364/AO.25.002715.

Bremner, S. P., M. Y. Levy, and C. Bo Honsberg. 2008. "Analysis of Tandem Solar Cell Efficiencies under AM1. 5G Spectrum Using a Rapid Flux Calculation Method." *Progress in Photovoltaics: Research and Applications* 16 (3): 225–233.

Fraunhofer Institute for Solar Energy Systems, 2019. "Photovoltaics Report." https://www.ise.fraunhofer.de/content/dam/ise/de/documents/publications/studies/Photovoltaics-Report.pdf.

Green, Martin A. 2008. "Self-Consistent Optical Parameters of Intrinsic Silicon at 300K Including Temperature Coefficients." *Solar Energy Materials and Solar Cells* 92 (11): 1305–1310. https://doi.org/10.1016/j.solmat.2008.06.009.

Holman, Z. C., A. Descoeudres, S. De Wolf, and C. Ballif. 2013. "Record Infrared Internal Quantum Efficiency in Silicon Heterojunction Solar Cells With Dielectric/Metal Rear Reflectors." *IEEE Journal of Photovoltaics* 3 (4): 1243–1249. https://doi.org/10.1109/JPHOTOV.2013.2276484.

McEvoy, A. J. 2003. "Photoelectrochemical Solar Cells." In *Practical Handbook of Photovoltaics: Fundamentals and Applications*. Boca Raton, Florida: CRC Press.

Nelson, Jenny. 2003. "Organic and Plastic Solar Cells." In *Practical Handbook of Photovoltaics: Fundamentals and Applications*, 484–511. Boca Raton, Florida: CRC Press.

National Renewable Energy Laboratory (NREL). 2019. "Best Research-Cell Efficiency." https://www.nrel.gov/pv/cell-efficiency.html.

National Renewable Energy Laboratory (NREL). n.d. "Reference Air Mass 1.5 Spectra." Accessed July 16, 2019. https://www.nrel.gov/grid/solar-resource/spectra-am1.5.html.

Panasonic. n.d. "N335 HIT® + Series." Accessed July 5, 2019. https://na.panasonic.com/us/energy-solutions/solar/hit-series/n335-hitr-series.

Philipps, Simon P., Andreas W. Bett, Kelsey Horowitz, and Sarah Kurtz. 2015. "Current Status of Concentrator Photovoltaic (CPV) Technology." NREL/TP–5J00-65130, 1351597. https://doi.org/10.2172/1351597.

Platzer, Michaela D. "U.S. Solar Photovoltaic Manufacturing: Industry Trends, Global Competition, Federal Support." Congressional Research Service, January 31, 2015.

Polman, Albert, Mark Knight, Erik C. Garnett, Bruno Ehrler, and Wim C. Sinke. 2016. "Photovoltaic Materials: Present Efficiencies and Future Challenges." *Science* 352 (6283). https://doi.org/10.1126/science.aad4424.

Rühle, Sven. 2016. "Tabulated Values of the Shockley–Queisser Limit for Single Junction Solar Cells." *Solar Energy* 130: 139–147.

3

First-Generation Solar Cells

The first photovoltaic (PV) cell was built and demonstrated by French physicist Edmond Becquerel in 1839. But, it was more than a century later in 1954 that PV technology took a leap forward in the United States when Bell Labs demonstrated the first solar cell to power an electronic device. In the late 1950s, these first-generation PV cells were produced commercially out of crystalline silicon (U.S. Department of Energy [DOE] n.d.) and similar structures continue to thrive commercially today. First-generation technologies based on crystalline silicon are now mature. At 93% of global market share (DOE n.d.), crystalline forms of silicon dominate traditional solar energy installations by a very wide margin. Crystalline silicon can take one of two forms in PV cells: monocrystalline or polycrystalline (Figure 3.1a and b). Amorphous silicon (Figure 3.1c) is a disorganized, noncrystalline form of silicon that can also be used to make commercially viable PV cells, but it is considered to be part of the second generation of solar cells.

3.1 Monocrystalline Silicon

While a single layer of silicon is able to generate current-carrying electrons and holes by absorbing photons of incoming light, a large majority of these electrons and holes will recombine with one another before they can be collected as useful electrical current. In order to reduce recombination, two different types of silicon are often fabricated to form a junction. Such a single junction consists of a single layer of p-type silicon fabricated adjacent to a single layer of n-type silicon. Both p-type and n-type layers consist of well-ordered, continuous, unbroken crystalline material formed and grown from a single seed of high-purity silicon. The p-type layer is doped with acceptor (hole-rich) atoms like boron and the n-type layer is doped with donor (electron-rich) atoms like phosphorous. The p-n junction consisting of these two layers is the active (absorber) layer of a first-generation, silicon-based PV cell, whose physical layers are often ordered as shown in Figure 3.2. The remaining layers in the PV cell structure (Table 3.1) are designed to provide an energetically favorable environment for holes to flow to the back of the cell and electrons to flow to the front of the PV cell.

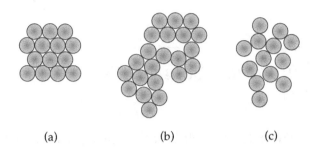

(a)   (b)   (c)

FIGURE 3.1
Various forms of silicon used in PV cells. The degree of order in silicon structure decreases from (a) monocrystalline silicon to (b) polycrystalline silicon to (c) amorphous silicon.

Light enters the PV cell through a transparent superstrate such as glass. Underneath the glass is a layer of antireflective coating that minimizes the loss of light by reflection. Silicon nitride is a common choice for this antireflective coating because it is highly compatible with silicon fabrication processes and its refractive index provides very good antireflective properties. Metal contacts (e.g., silver, aluminum) fabricated onto the front-side or top of the PV cell make connections to other PV cells and to external circuits. These front-side contacts are patterned in such a way as to cover only a small portion of the PV cell surface in order to allow most light to pass through into the working (active) p-n junction. The function of the front-side contacts is to collect the electrons that are generated in the active layers of the cell. The active layers are fabricated underneath the antireflective coating and front-side contacts and consist first of an n-type silicon layer and then the p-type layer, although these layers may be swapped or inverted in some PV cell structures. Throughout the entire

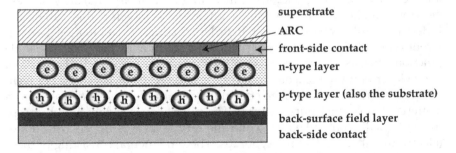

superstrate
ARC
front-side contact
n-type layer
p-type layer (also the substrate)
back-surface field layer
back-side contact

FIGURE 3.2
Silicon PV cell structure. The two doped (p-type and n-type) silicon layers absorb light in the silicon PV cell. Recombination is minimized by the action of the built-in potential and electric field at the p-n junction. The front-side contact collects free electrons associated with light-generated electron-hole pairs and the back-side contact collects free holes. (From Chen 2015.)

TABLE 3.1

Typical Composition of a Crystalline Silicon PV Cell

Layer	Typical Material	Function
Superstrate	Glass	Transparent; provides structural stability and protection from the environment
Antireflective coating (ARC)	Silicon nitride (Si_3N_4)	Minimizes reflection of light away from the PV cell
Front-side contact	Silver (Ag) or aluminum (Al)	Collects electrons
n-type layer p-n junction p-type layer	Monocrystalline or polycrystalline silicon	Transports electrons Generates electron-hole pairs Transports holes
Back-surface field layer	Heavily doped p^{++} silicon	Blocks electrons from back-side contact
Back-side contact	Silver (Ag) or aluminum (Al)	Collects holes

Source: Chen (2015).

bilayer (p-n) junction, light is absorbed as photons, generating electron-hole pairs. Within a certain distance from the p-n junction, however, these electron-hole pairs remain separated long enough that most can travel to the front-side and back-side contacts and then to an external circuit for conversion to useful electricity.

Underneath the p-type active layer and near the bottom of the PV cell are two more layers. The first layer acts as a back-surface field layer that passivates the back of the PV cell and prevents minority electrons in the p-type layer from flowing into the bottom of the cell where they would be otherwise free to recombine with holes and subsequently reduce collection efficiency and overall electrical current. And finally, under the back-surface field layer is a back-side contact made of aluminum, silver, or similar conductor that is energetically favorable for collecting the holes that were generated in the active layer of the cell (Chen 2015).

As described in Chapter 2, electrons and holes that are generated within the depletion region at the junction of p-type and n-type layers and within one diffusion length of this depletion region are less likely to recombine and more likely to reach the contacts on both sides of the PV cell. Once at the contacts, they can be collected in the circuit external to the PV cell as useful electrical current and energy. The built-in potential of the p-n junction provides a strong electric field force to sweep electrons and holes in opposite directions and thereby prevent recombination. The end result of this process is that up to 26% of incoming light energy (National Renewable Energy Laboratory [NREL] 2019) can effectively and efficiently be converted to electrical energy, leading to efficiencies in commercially available PV modules that range between 15% and 20% (Energy Informative 2013).

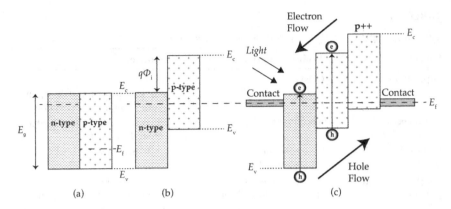

FIGURE 3.3
Energy band structure in silicon PV cells. (a) p-n junction prior to drift and diffusion of electrons and holes; (b) p-n junction at equilibrium in dark (i.e., zero light) conditions; (c) p-n junction integrated into the PV cell under illuminated conditions at equilibrium (i.e., no current flow). At equilibrium, the Fermi level E_f is flat across all layers in the PV cell structure. In practical devices, conduction and valence bands bend near the interfaces between different layers rather than experiencing the abrupt changes suggested by the simplified diagram above.

A convenient way to visualize the movement of electrons and holes in this first-generation PV cell is to look at the cell from an energy point of view by constructing the energy band diagrams of the various layers that make up the cell. The energies of the electrically functional layers of the PV cell structure in Figure 3.2 are shown in Figure 3.3. The gap between the valence band and conduction band edges is the energy bandgap (E_g) of silicon and is approximately 1.14 eV at room temperature. The Fermi energy level (E_f) is also shown for all layers in the PV structure. E_f corresponds to the energy at which the probability of finding a free electron or hole at a lower energy level is 0.5. E_f also reflects the balance of free electrons and free holes in a semiconductor. In an intrinsic semiconductor, E_f lies in the middle of the energy gap, reflecting that the number of free electrons in the conduction band is about equal to the number of free holes in the valence band. In the p-type layer, the Fermi level (E_f) energy lies close to the valence band, reflecting the fact that the material is dominated by free holes in the valence band and that free electrons in the conduction band are in the minority. In the n-type layer, the situation is the opposite. E_f lies close to the conduction band, reflecting the fact that an n-type semiconductor is dominated by free electrons in the conduction band. In a metal, the conduction and valence bands overlap and the Fermi level has an energy equal to the maximum energy an electron in that metal can have at absolute zero temperature. In some heavily doped semiconductors, the Fermi level can also lie inside the valence or conduction band.

When layers of different materials are joined together, equilibrium is reached when the Fermi levels align with one another. When p-type and n-type layers are first joined, the Fermi energy levels are different (Figure 3.3a), but equilibrium causes the two Fermi levels to align (Figure 3.3b). To facilitate equilibrium, electrons from the electron-rich n-type region diffuse to the p-type region (and vice versa for holes). As a result of diffusion, the p-side of the junction becomes negatively charged and the n-side of the junction becomes positively charged. Since negative charges are attracted to positive charges, drift soon counteracts diffusion and electrons on the negatively charged p-side of the junction drift back to the n-side of the junction based on this attraction. The balance between drift and diffusion occurs in equilibrium at the built-in potential of the p-n junction (shown as $q\phi_i$ in Figure 3.3b). The built-in potential is a function both of the material and the doping on either side of the junction, as described in detail in Chapter 2. This built-in potential creates an energy hill where electrons in the conduction band can flow "down" the hill and holes in the valence band can "float" up the hill. The value of $q\phi_i$ in a silicon p-n junction varies from approximately 0.56 eV in a lightly doped device (for one dopant atom per 100 million silicon atoms) to 1.04 eV in a heavily doped device (for one dopant atom per 10,000 silicon atoms).

For the remainder of the PV cell, equilibrium requires that the Fermi levels be at the same energies across all layers in the material from front-side contact to back-side contact (Figure 3.3c). Once the Fermi levels of all five current-carrying layers in the PV cells reach the same energy, it can be seen that the entire PV cell facilitates the flow of electrons "down" energy hills from the heavily doped, p^{++} silicon back-field surface layer to the front-side contact. Likewise, the PV cell facilitates the floating of holes "up" energy hills from the n-type layer to the back-side contact. As light generated carriers flows through the complete PV cell, the height of the energy hill decreases from its initial built-in potential ($q\phi_i$) to a flat energy band when the open circuit voltage of the PV cell is reached.

Bilayer, p-n junction PV cells can be connected in series to increase voltage or in parallel to increase the total photocurrent of a PV cell array. An array of PV cells in combination with the circuits required to collect, maximize, and condition current make up a solar panel. In traditional rooftop and field-based solar systems, multiple panels are required to generate a desired power level. Solar panels based on monocrystalline silicon are widespread. But, despite the relatively high efficiency of a single silicon PV cell, monocrystalline silicon is imperfect and not without its problems. For example, because of the complexity and precision of the manufacturing process required to produce the high-purity expected of monocrystalline silicon, considerable energy is consumed during the actual production of solar panels. In fact, a monocrystalline silicon solar panel can take up to 25% of its life span to compensate for the amount of energy it takes to produce the panel (Dai Pra, Dias, and Kieling 2015). The growth process for monocrystalline silicon is also wasteful. Monocrystalline silicon is grown in a tube-like shape and a great deal of material waste is generated when the silicon is cut and shaped into rectangular cells for commercial panels. Monocrystalline silicon PV cells are also considerably thicker than other PV cell technologies and are both rigid and brittle. For wearable solar cell systems that are subject to bending and flexing, these physical properties make monocrystalline silicon highly vulnerable to cracking and permanent damage.

In addition to being wasteful and fragile, monocrystalline silicon PV cells are prone to performing poorly when subject to shading, soiling, or other irregularities in light coverage on the surface of a solar panel. In fact, on rooftop installations, the effect of dust and other soiling has been shown to decrease monocrystalline PV performance by up to 50% (Sulaiman and Hussain 2011). Because of their mobile and portable nature, wearable solar cell systems are subject to even more frequent bouts of shading and other irregularities than rooftop installations, thus amplifying these liabilities of monocrystalline silicon even further.

Wearable solar cell systems do offer an advantage to monocrystalline silicon PV cells. Because of their very nature, these wearable systems will not see as high an operating temperature as solar panels baking on a rooftop in direct sunlight nor as cold an operating temperature as a solar panel waiting patiently through a blizzard. Given that both high and low temperatures in these PV cells can significantly degrade the overall performance, the more narrow temperature ranges experienced by wearables result in more stable performance.

Of all the downsides of monocrystalline silicon for use in wearable solar cell systems, the most important is likely to be the cost. Monocrystalline silicon is the most expensive among silicon solar cells, at about 75 cents per Watt (of power production capacity) for stationary modules (Energy Informative 2013).

3.2 Polycrystalline Silicon

A lower-cost alternative to monocrystalline silicon is polycrystalline silicon. Polycrystalline silicon is often used as a synonym for multicrystalline silicon, although the term multicrystalline is likely to describe silicon with much larger crystals (on the order of a millimeter). Both polycrystalline and multicrystalline silicon (Figure 3.1b) are less ordered than monocrystalline silicon (Figure 3.1a). But, polycrystalline silicon, like monocrystalline silicon, begins with a single seed of crystal upon which the remaining silicon grows. Unlike monocrystalline silicon, the seed is not pulled upward during the growth process, and the resulting silicon has edges and grains that lead to a more disorganized although still crystalline structure. Edges and grains also disrupt the flow of electrons and decrease the energy conversion efficiency. PV cells made from polycrystalline silicon are fabricated in a very similar way to PV cells made from monocrystalline silicon (Figure 3.2) and have similar energy band structures (Figure 3.3) that control the flow of electrons and holes through the cell.

Historically, the lower quality and irregular structure of polycrystalline silicon have caused polycrystalline PV cells to lag behind monocrystalline silicon in efficiency. However, advances in manufacturing for polysilicon PV cells have reduced the performance gap enough so that the difference in best research cell efficiencies between the two types of PV cells is only about 4% (NREL 2019). Once assembled into fully functioning solar panels, polycrystalline silicon panels have an efficiency between 13% and 16%, whereas monocrystalline solar panels are commercially available at efficiencies between 15% and 20% (Energy Informative 2013). Polysilicon PV cells cost less to manufacture and generate less waste, which offsets their reduced efficiency. In choosing between the two types of PV cells, the bottom line often comes down to the space available to place and install solar panels. If space is not an issue, the lower costs of polysilicon solar cells and systems make them a better choice. If space is at a premium, monocrystalline silicon often wins out, despite its increased cost.

3.3 Amorphous Silicon

Crystalline forms of silicon, including monocrystalline, polycrystalline, and multicrystalline, together make up the first generation of solar cells. Amorphous silicon is a noncrystalline, disordered form of silicon (Figure 3.1c) that is technically considered a member of the second generation of solar cells. Because they are still fabricated from silicon as the underlying active material, however, amorphous silicon PV cells are discussed here. Despite the disorder in its underlying structure, amorphous silicon can be fabricated into useful PV cells (Figure 3.4) using a similar structure and composition (Table 3.2) as crystalline PV cells.

Just as with crystalline silicon PV cells, amorphous silicon PV cells are covered with a superstrate that is typically made from glass to provide structural stability and protect the PV cell from the ambient environment. Directly underneath the glass are front-side contacts which collect holes generated by incident light. The next layer is an antireflective, transparent conducting electrode often made of fluorine-doped tin oxide that provides a low-resistance contact between the active layers of the PV cell and the front-side contact and also reduces surface recombination of free holes. Under the electrode is the active area of the cell, consisting of a p-type silicon layer, an insulating layer, and an n-type silicon layer, all made of hydrogenated amorphous silicon. Unlike crystalline silicon PV cells, the p-type and n-type layers are very thin, on the order of 100s of nanometers (Qarony et al. 2017),

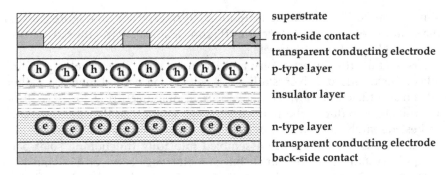

superstrate
front-side contact
transparent conducting electrode
p-type layer

insulator layer

n-type layer
transparent conducting electrode
back-side contact

FIGURE 3.4
Amorphous PV cell structure. Amorphous PV cells are often inverted from silicon PV structures with the p-type layer closer to the surface of the cell. Amorphous PV cells are also made with a p-i-n junction rather than a p-n junction where the "i" refers to an insulator. The insulator layer between the p-type and n-type layers increases the number of electron-hole pairs generated by incoming light to the benefit of the overall efficiency of the PV cell. The transparent conducting electrode is used as a low resistance, transparent contact between the front-side contact and the p-type layer and again, between the back-side contact and the n-type layer. Both electrodes are degenerate—so heavily doped that they act like metals. Both electrodes also have large bandgaps so that they are largely transparent to incoming light. (From Qarony et al. 2017.)

and therefore they have limited capacity to produce electron-hole pairs in response to incoming light. The insulator layer provides greater volume over which light can generate these electron-hole pairs with minimal recombination, thus improving the energy-conversion efficiency of the overall structure. Aside from this insulator layer that increases the number of

TABLE 3.2

Typical Composition of an Amorphous Silicon PV Cell

Layer	Material	Function
Superstrate	Glass	Transparent; provides structural stability and protection from the environment
Front-side contact	Silver	Collects holes
Transparent conducting electrode	Fluorine-doped tin oxide (FTO)	Antireflective coating (ARC); low resistance contact to adjacent layers
p-type layer	a-SiC:H	Transports holes
Insulator layer[a]	a-Si:H	Generates electron-hole pairs
n-type layer[a]	a-Si:H	Transports electrons
Transparent conducting electrode	Aluminum-doped zinc oxide (AZO)	Blocks electrons in back-side contact from entering the PV cell
Back-side contact	Silver (Ag)	Collects electrons

Source: Qarony et al. (2017).
[a] Hydrogenated amorphous silicon is used in these layers for its low-defect density.

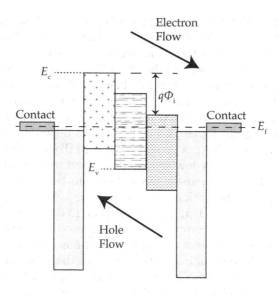

FIGURE 3.5

Energy band structure of amorphous PV cell. In the amorphous PV cell structure shown in Figure 3.4, electrons flow toward the back-side of the cell (to the right in the above figure) and holes float toward the top of the cell (to the left). The heavily doped transparent conducting electrode acts like a metal, providing a low resistance contact for holes in the p-type layer to successfully travel to the front-side contact for collection. The transparent conducting electrode near the bottom of the cell acts in a similar manner to promote the transport of electrons to the back-side contact.

electron-hole pairs generated from incoming light, the p-n junction structure functions in the same way as that for crystalline PV cells.

The amorphous silicon PV cell shown in Figure 3.4 is inverted (i.e., layers are in the opposite order from front to back of the cell) from the crystalline silicon PV cell (Figure 3.2). As a result, in energy terms (Figure 3.5), light-generated electrons flow down the energy hill toward the back of the cell, while holes float up the energy hill toward the front of the cell. The transparent conducting electrode at both the front and the back of the PV cell is doped in such a way that the Fermi level is above the conduction band edge to facilitate the proper flow of electrons and holes for best energy conversion efficiency.

Most amorphous silicon is hydrogenated (a-Si:H) for use in PV cells and while lacking a formal energy bandgap has a mobility gap whose value ranges between 1.7 and 1.9 eV and performs roughly the same function as the energy bandgap in crystalline silicon structures. Because of dangling bonds and other irregularities and defects in the disordered amorphous structure, amorphous silicon has a much lower hole mobility than crystalline silicon. While hole mobilities in crystalline silicon are often over

100 cm^2/V-sec (El-Cat n.d.), the hole mobility of amorphous silicon is on the order of 0.01 cm^2/V-sec (Schiff 2006). The inherently poor hole mobility is a major contributor to the reduced efficiencies of amorphous silicon PV cells that have only reached about 14% (NREL 2019). Because of these low efficiencies, amorphous silicon has historically been used only for low-power solar-powered devices like calculators. But, the fact that amorphous silicon can be made into very thin PV cells drastically reduces its cost and allows it to be adapted to irregular topologies because of its flexibility and bendability. Amorphous silicon is also not grown from seed like crystalline silicon, but it can be deposited at very low temperatures and over large areas, thereby enabling deposition onto plastic, fabrication into large individual PV cells, reduced waste, and integration into inexpensive roll-to-roll manufacturing techniques. All of these advantages contribute to dramatically reduced PV cell and solar panel costs compared to crystalline silicon. Furthermore, thin film PV cells such as those made with amorphous silicon leak less current and perform much better at low-light conditions than thicker PV cells based on crystalline silicon. These advantages can counteract disadvantages of reduced efficiency and reduced lifetime for a variety of applications including wearable solar cell systems.

3.4 Summary

By far, silicon is the most popular choice for commercial PV cells and solar panels. With high efficiencies (>25%) and mature fabrication technology, crystalline silicon is well poised to support stationary and general-purpose solar installations for years to come. For wearable solar cell systems where a rigid structure is tolerable, crystalline silicon is also a good choice. When flexibility is essential, lower-efficiency, lower-cost PV cells made with amorphous silicon can be wrapped around just about anything including smart clothing and a wide number of accessories worn on the human body. When considering all types of silicon PV cells together, it may be hard to believe that there is an application for solar cells that needs something beyond silicon. But, second- and third-generation solar cells strive to tell the story otherwise.

References

Chen, Jiahe. 2015. "Recent Developments on Silicon Based Solar Cell Technologies and Their Industrial Applications." *Energy Efficiency Improvements in Smart Grid Components.* https://doi.org/10.5772/59171.

Dai Pra, Lea Beatriz, Joao Batista Dias, and Amanda Goncalves Kieling. 2015. "Comparison between the Energy Required for Production." *Journal of Energy and Power Engineering* 9: 592–597. https://doi.org/10.17265/1934-8975/2015.06.011.

El-Cat. n.d. "Properties of Silicon." Accessed May 11, 2019. https://www.el-cat.com/silicon-properties.htm.

Energy Informative. 2013. "Solar Cell Comparison Chart – Mono-, Polycrystalline and Thin Film." https://energyinformative.org/solar-cell-comparison-chart-mono-polycrystalline-thin-film/.

National Renewable Energy Laboratory (NREL). 2019. "Best Research-Cell Efficiency." https://www.nrel.gov/pv/cell-efficiency.html.

Qarony, Wayesh, Mohammad I. Hossain, M. Khalid Hossain, M. Jalal Uddin, A. Haque, A. R. Saad, and Yuen Hong Tsang. 2017. "Efficient Amorphous Silicon Solar Cells: Characterization, Optimization, and Optical Loss Analysis." *Results in Physics* 7: 4287–4293.

Schiff, Eric A. 2006. "Hole Mobilities and the Physics of Amorphous Silicon Solar Cells." *Journal of Non-Crystalline Solids* 352 (9–20): 1087–1092.

Sulaiman, Shaharin A., and Haizatul H. Hussain. 2011. "Effects of Dust on the Performance of PV Panels." *International Journal of Mechanical and Mechatronics Engineering* 5 (10): 6.

U.S. Department of Energy (DOE). n.d. "History of Solar." https://www1.eere.energy.gov/solar/pdfs/solar_timeline.pdf.

4

Second-Generation Solar Cells

Second-generation photovoltaics (PVs) refer to noncrystalline silicon or non-silicon inorganic semiconductors fabricated in similar structures to crystalline silicon PV cells, albeit in much thinner films. At almost 7% of global market share (Solar Central n.d.), these PV cells make up most of the remaining global market share for solar panels. The three most promising materials of the second-generation options discussed in this chapter are gallium arsenide (GaAs), cadmium telluride (CdTe), and copper indium gallium selenide (CIGS). Another second-generation material, amorphous silicon, can also compete with first-generation solar cells and was addressed in Chapter 3 with other silicon-based solar cells.

4.1 Gallium Arsenide

Like silicon, gallium arsenide (GaAs) is an inorganic semiconductor that can be fabricated into a PV cell by joining a p-type (acceptor) doped GaAs layer with an n-type (donor) doped GaAs layer. The structure of a GaAs PV cell (Figure 4.1, Table 4.1) is very similar to the silicon PV cell structure.

The energy band structure of the GaAs PV cell is also similar to that of the corresponding crystalline silicon PV cell (Figure 3.3) except that GaAs has a wider bandgap than silicon (1.43 eV) and subsequently, greater built-in potentials, leading to a larger open circuit voltage in the PV cell.

Some modern GaAs technologies dispense with the substrate layer in order to increase solar cell flexibility and reduce costs (Alta Devices n.d.). Furthermore, GaAs and other thin-film structures often use a window layer near the surface of the PV cell. Since a large number of photons of light are absorbed near the surface of the cell, the window layer prevents electron-hole pairs that result from the absorption of photons from reaching the surface of the cell where the probability of recombining and subsequent loss of electrical energy is high. The window layer usually consists of one or more transparent metal oxides that have large bandgaps whose function is to prevent electrons and holes from migrating to the PV cell surface.

While the GaAs PV cell operates very similarly to silicon-based PV cells, there are a few major distinctions between the two materials. The larger bandgap of GaAs means that the longest wavelength to which GaAs

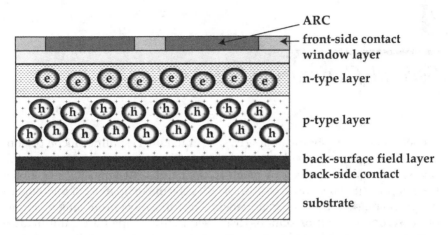

FIGURE 4.1

Gallium arsenide (GaAs) PV cell structure. The GaAs PV cell structure is similar to that of a crystalline silicon PV cell except for the addition of a window layer near the top of the cell. The window layer prevents electron-hole pairs from recombining at the surface. This is especially important for a direct bandgap semiconductor like GaAs because large numbers of photons are absorbed near the surface of the PV cell. At the bottom of the cell, silicon can be used as the substrate for structural stability to reduce PV cell cost. (From Xiao et al. 2018.)

responds is shorter than that for silicon, and therefore GaAs responds less to infrared light. When responding to sunlight, this difference in bandgaps means that GaAs has a slightly lower maximum or ultimate efficiency than silicon (Belghachi 2015). And, GaAs is also a direct bandgap semiconductor, while silicon is an indirect bandgap material. An indirect bandgap means that for an electron to travel from the highest energy level in the valence

TABLE 4.1

Typical Composition of a GaAs PV Cell

Layer	Typical Materials	Function
Front-side contact	Ni/Au/Ge/Ni/Au	Collects electrons
Antireflective coating (ARC)	Silicon nitride (Si_3N_4)	Minimizes reflection of light away from the PV cell
Window layer	n-type indium gallium phosphide (InGaP)	Reduces recombination
n-type layer	Gallium arsenide (GaAs)	Transports electrons
p-type layer		Generates electron-hole pairs Transports holes
Back-surface field layer	p-type indium gallium phosphide (InGaP)	Blocks electrons from back-side contact
Back-side contact	Platinum/gold (Pt/Au)	Collects holes
Substrate	Silicon (Si)	Provides structural stability

Source: Xiao et al. (2018).

band (E_v) to the lowest energy level in the conduction band (E_c), the electron must experience both a change in energy and a change in momentum. In contrast, a direct bandgap semiconductor requires only that an electron experiences a change in energy in order to move from E_v to E_c and vice versa from E_c to E_v. The direct bandgap means that GaAs has a shorter (minority carrier) diffusion length than silicon and as a result, electrons travel much shorter distances before recombining. But, GaAs also has stronger absorption characteristics than silicon, thereby enabling much thinner devices to absorb the same total amount of light as an equivalent silicon-based PV cell. The net effect of these differences between GaAs and silicon-based PV cells is that demonstrated research cell efficiencies for GaAs are as high as 29.1%, while monocrystalline silicon falls much shorter of its ultimate efficiency at 26.1% (National Renewable Energy Laboratory [NREL] 2019).

Since GaAs absorbs more light than its silicon counterparts, it can be made into films that are sufficiently thin to be flexible, not easily fractured or broken, and suitable for wearable solar cell systems. Unlike silicon, GaAs also demonstrates good low-light performance and is capable of generating useful electrical energy even in the artificially lit environments to which wearable solar cell systems will be frequently exposed. Exceptional low-light performance comes about as a result of lower leakage currents in GaAs that reduce power losses and stabilize energy conversion efficiencies at lower light levels. Low-light performance and cell flexibility are major advantages for GaAs PV cells compared to first-generation silicon PV cells, particularly in wearable solar cell systems, where irregular topologies and low light are more the norm than the exception.

Although less relevant to wearable solar cell systems that are more likely than traditional systems to be operated in sheltered environments, GaAs is resistant to ultraviolet light, moisture damage, and performance degradation at high temperatures. This robustness makes PV cells made with GaAs well suited to aerospace and space applications, whether manned or unmanned, wearable or not, long-term or temporary.

Unfortunately, in all but the most high-performance applications, these numerous benefits of GaAs PV cells have historically been overshadowed by their high cost. Cost has limited these PV cells to applications where higher costs are tolerable for the performance and power provided by GaAs PV technology. Recently, however, advances in the fabrication of GaAs PV cells have made increasingly thin films of one micron commercially viable (Wilkins 2018). Such small geometries have a major impact in reducing overall costs and enabling very lightweight solar modules that may be attractive for traditional rooftop and other stationary installations. Increased flexibility and reduced weight are also attractive for wearable systems, although the presence of arsenic in GaAs PV cells may limit their use in wearable energy harvesting systems. In certain forms, arsenic is carcinogenic to humans (Ratnaike 2003) and while GaAs has not been definitively tied to cancer, its carcinogenic potential is still being explored (Bomhard et al. 2013).

4.2 Cadmium Telluride

Like GaAs, CdTe is also a direct bandgap semiconductor that enables it to absorb more light in a thinner film than silicon, consistent with Beer's law (Equation 2.5). At a wavelength of 360 nm, silicon, GaAs, and CdTe all have an absorption coefficient (α) on the order 10^6 cm^{-1}, which means that they all absorb about the same amount of violet light for a given film thickness. By 560 nm (green light), silicon has lost two orders of magnitude in the absorption coefficient ($\alpha = 10^4$ cm^{-1}), while CdTe and GaAs have only lost one order of magnitude ($\alpha = 10^5$ cm^{-1}). And, by 800 nm, the gap in absorption widens even further (PV Education 2019). Since the spectrum of sunlight as well as many artificial light sources contains significant percentages of energy in the green and longer wavelengths, this difference in absorption coefficients results in far less light absorption for the same geometry in silicon compared to both CdTe and GaAs. But, unlike GaAs, costs for CdTe solar systems are comparable to crystalline silicon solar systems, thus providing CdTe ample opportunity to break into the solar energy market. In 2017, CdTe was more successful than any other PV technology in going head-to-head with silicon, capturing 2.4% of the global energy produced from PV-based solar energy systems (Fraunhofer Institute for Solar Energy Systems 2019).

A CdTe PV cell has a similar structure to all other semiconductor-based PV cells (Figure 4.2, Table 4.2) except that the p-n junction is heterogenous (made of two different materials) rather than homogenous.

The active (light absorbing) layer of the CdTe PV cell consists of a thin n-type layer of cadmium sulfide (CdS) that is approximately 0.2 μm thick and a much thicker layer of p-type CdTe. As with other semiconductor-based

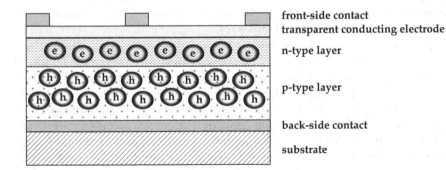

FIGURE 4.2
Cadmium telluride (CdTe) PV cell structure. The transparent conducting electrode acts as a low-resistance interface between the n-type layer and the front-side contact to enable more efficient collection of free electrons generated by the absorption of photons. Free holes are collected at the back-side of the cell. (From Abdullah, Razooqi, and Al-Ajili 2013.)

TABLE 4.2

Typical Composition of a CdTe PV Cell

Layer	Material	Function
Front-side contact	Aluminum (Al)	Collects electrons
Transparent conducting electrode	Tin oxide (SnO_2)	Antireflective coating; low-resistance contact to adjacent layers
n-type layer	Cadmium sulfide (CdS)	Transports electrons and acts as window layer
p-type layer	CdTe	Transports holes
Back-side contact	Gold (Au)	Collects holes
Substrate	Glass	Provides structural stability

Source: Abdullah, Razooqi, and Al-Ajili (2013).

bilayer junctions, the n-type layer transports electrons and the p-type layer transports holes generated by light irradiating the p-n junction. On top of the n-type layer is a transparent conducting electrode that acts in combination with the CdS layer to reduce recombination and improve collection efficiencies. This transparent conductive electrode also acts (a) as an antireflective coating; (b) to collect electrons; and (c) to transmit light to the underlying p-n structure. Electrical contact to the transparent conducting electrode is made through a patterned front-side contact layer of metal that covers only a small portion of the surface of the PV cell.

On the back-side of the PV cell, a back-side contact collects free holes that are transported through the p-side of the active p-n junction and the entire cell sits on a substrate such as glass that provides structural stability (Abdullah, Razooqi, and Al-Ajili 2013).

The energy band structure of a CdTe PV cell (Figure 4.3) is different from silicon and GaAs PV cells because the active area is a heterojunction consisting of two different semiconducting materials (CdTe and CdS). The bandgap of CdS is about 2.4 eV, much wider than silicon, GaAs, and CdTe, while the bandgap of CdTe is closer to that of GaAs at 1.5 eV. Like GaAs and silicon, the bandgap of CdTe is well matched to sunlight, while CdS has a much wider bandgap and stops absorbing light at much shorter wavelengths than other semiconductors.

Because of their direct bandgap, CdTe PV cells have excellent light absorption characteristics and are well suited for thin films. Research cell efficiencies of CdTe thin films have reached 22.1% (NREL 2019), while solar module system efficiencies are as high as 16.1% (Rix et al. 2015). Combined with low cost, these characteristics make CdTe a popular choice for conventional solar panels, ranking second behind crystalline silicon. And, like other thin-film technologies, CdTe PV cells are flexible and well suited to the irregular and changing topologies of wearable solar cell systems.

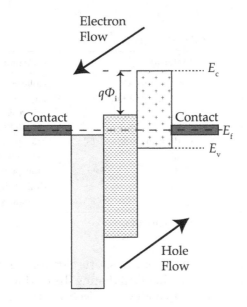

FIGURE 4.3

Energy band structure of CdTe PV cells. The energy band structure corresponds to the physical structure shown in Figure 4.2. In energy terms, electrons flow down the energy hill toward the front-side of the cell (to the left in the above figure) and holes float up the energy hill toward the back of the cell. At equilibrium, the Fermi levels align and conduction and valence bands bend at the interfaces (not shown). The transparent conducting electrode at the top of the cell is heavily doped and acts as a low-resistance contact between the n-type layer and the front-side contact.

CdTe films compete well with both amorphous silicon thin films and conventional crystalline PV cells and can be made at comparable or lower cost than monocrystalline silicon PV cells (Horowitz et al. 2017). CdTe is also more stable at higher temperatures of operation than its silicon cousins. CdTe has the added benefit of reduced environmental impact during manufacturing because these PV cells use less water and generate fewer greenhouse gases than many other PV cell technologies. However, tellurium is a very rare element that is extracted from the copper purification process and requires very large amounts of copper to produce correspondingly small amounts of tellurium. For this reason alone, solar cell systems made with CdTe are not likely to be sustainable over the long term. Complicating matters further, cadmium is a toxic heavy metal with serious impacts on human kidneys and respiratory, reproductive, and skeletal systems (Godt et al. 2006). The toxicity of cadmium has serious environmental and public health implications at the end of the life cycle of traditional solar panels from CdTe. Careful and proper disposal of CdTe panels is critical, but in wearable solar cell systems, there is also the potential for cadmium exposure during the use of CdTe PV cells. Thus, the risk to public health may be simply too high for CdTe to take hold in wearable markets.

4.3 Copper Indium Gallium Selenide

Reducing cadmium and eliminating tellurium is a strategic approach to bypassing the significant negative impacts on natural resources, environment, and public health associated with CdTe systems. CIGS fills the bill. CIGS thin films are the last of the four major players (amorphous silicon, GaAs, CdTe, and CIGS) among second-generation, inorganic PV cells. The best research cell efficiencies of CIGS cells are competitive with other first- and second-generation solar cell technologies, at 23.4%, compared to 22.1% for CdTe and 26.1% for crystalline silicon-based PV cells (NREL 2019). The physical structure of the CIGS PV cell is very similar to the CdTe cell (Figure 4.2) as is the energy band structure (Figure 4.3). The composition of a typical CIGS cell is described in Table 4.3.

Like CdTe and GaAs, CIGS PV cells are made from direct bandgap semiconductors which allow for greater light absorption and subsequently, much thinner devices overall. An added benefit of CIGS films is that the energy bandgap of the CIGS layer can be adjusted or tuned to values between 1 and 2.4 eV by changing the ratios of the CIGS elements (copper, indium, gallium, selenium) to maximize the absorption of light (Polman et al. 2016). At lower bandgaps, CIGS thin films have the highest research cell efficiencies among the thin films at 22.9% (NREL 2019), but difficulties in transitioning from successful fabrication in the laboratory to mass production have kept efficiencies of both small and large CIGS solar modules below 20% (Powalla et al. 2017).

While CIGS still uses a layer containing cadmium (i.e., the n-type CdS layer), it is much thinner than the p-type CdTe layer in CdTe PV cells. As a

TABLE 4.3

Typical Composition of a CIGS PV Cell

Layer	Material	Function
Window layer	Aluminum zinc oxide (AZO)	Transparent; reduces recombination and current loss
Transparent conducting electrode	Zinc oxide (ZnO)	Antireflective; Collects electrons
n-type layer	n-type cadmium sulfide (CdS)	Transports electrons
p-type layer	p-type copper indium gallium selenide (CIGS)	Transports holes
Back-side contact	Molybdenum (Mo)	Collects holes; reflects light back into CIGS
Substrate	Soda lime glass, metal foil, or plastic	Increases open circuit voltage of cell; provides structural stability

Source: NREL (n.d.) and Parisi et al. (2015).

result, the total amount of cadmium used in these cells is greatly reduced as is the risk to human health during use and to ecosystem health during disposal and recycling. The fact that CIGS cells do not use the rare earth element tellurium is another win in the environment column for CIGS over CdTe.

Similar to CdTe, the performance of CIGS cell is also stable with increasing temperature, unlike silicon-based solar cells. Although temperature swings among wearable solar cell systems are likely to be smaller than those experienced by outdoor solar installations, temperature stability is nevertheless a desirable characteristic to maintain consistent and predictable power output.

4.4 Summary

While silicon remains the most popular choice for commercial PV cells and solar panels, second-generation thin films can compete in terms of the efficiency by which sunlight is converted into usable electricity. With high research cell efficiencies (29%, 22%, and 23% for GaAs, CdTe, and CIGS, respectively), second-generation, non-silicon solar cells offer comparable performance to silicon with added benefits. But, despite competitive performance, thin-film technologies are struggling to increase market share in traditional solar panels and modules in large part because of the maturity and inertia of silicon solar cell technology. In wearable solar cell systems, however, where flexibility and low-light performance can be as important as efficiency, thin-film inorganic semiconductor PV cells are particularly attractive. The low cost and ease of manufacturing of several of these second-generation technologies add to their overall appeal for wearable systems.

References

Abdullah, Rasha A., A. Razooqi Mohammed, and Al-Ajili Adwan N. H. 2013. "Characterization of the Energy Band Diagram of Fabricated SnO2/CdS/CdTe Solar Cells." *World Academy of Science, Engineering and Technology* 79: 118–122.

Alta Devices. n.d. "World's Most Efficient, Thin and Flexible Solar Technology." *Alta Devices* (blog). Accessed May 22, 2019. https://www.altadevices.com/technology/.

Belghachi, Abderrahmane. 2015. "Theoretical Calculation of the Efficiency Limit for Solar Cells." In *Solar Cells - New Approaches and Reviews*, edited by Leonid A. Kosyachenko. InTech. https://doi.org/10.5772/58914.

Bomhard, Ernst M., Heinz-Peter Gelbke, Hermann Schenk, Gary M. Williams, and Samuel M. Cohen. 2013. "Evaluation of the Carcinogenicity of Gallium Arsenide." *Critical Reviews in Toxicology* 43 (5): 436–466. https://doi.org/10.3109/10408444.2013.792329.

Fraunhofer Institute for Solar Energy Systems. 2019. "Photovoltaics Report." https://www.ise.fraunhofer.de/content/dam/ise/de/documents/publications/studies/Photovoltaics-Report.pdf.

Godt, Johannes, Franziska Scheidig, Christian Grosse-Siestrup, Vera Esche, Paul Brandenburg, Andrea Reich, and David A. Groneberg. 2006. "The Toxicity of Cadmium and Resulting Hazards for Human Health." *Journal of Occupational Medicine and Toxicology (London, England)* 1 (September): 22. https://doi.org/10.1186/1745-6673-1-22.

Horowitz, Kelsey A. W., Ran Fu, Tim Silverman, Mike Woodhouse, Xingshu Sun, and Mohammed A. Alam. 2017. "An Analysis of the Cost and Performance of Photovoltaic Systems as a Function of Module Area." NREL/TP–6A20-67006, 1351153. https://doi.org/10.2172/1351153.

National Renewable Energy Laboratory (NREL). 2019. "Best Research-Cell Efficiency." https://www.nrel.gov/pv/cell-efficiency.html.

National Renewable Energy Laboratory (NREL). n.d. "Copper Indium Gallium Diselenide Solar Cells | Photovoltaic Research | NREL." Accessed May 24, 2019. https://www.nrel.gov/pv/copper-indium-gallium-diselenide-solar-cells.html.

Parisi, Antonino, Riccardo Pernice, Vincenzo Rocca, Luciano Curcio, Salvatore Stivala, Alfonso C. Cino, Giovanni Cipriani, Vincenzo Di Dio, Giuseppe Ricco Galluzzo, and Rosario Miceli. 2015. "Graded Carrier Concentration Absorber Profile for High Efficiency CIGS Solar Cells." *International Journal of Photoenergy.* https://doi.org/10.1155/2015/410549.

Polman, Albert, Mark Knight, Erik C. Garnett, Bruno Ehrler, and Wim C. Sinke. 2016. "Photovoltaic Materials: Present Efficiencies and Future Challenges." *Science* 352 (6283). https://doi.org/10.1126/science.aad4424.

Powalla, Michael, Stefan Paetel, Dimitrios Hariskos, Roland Wuerz, Friedrich Kessler, Peter Lechner, Wiltraud Wischmann, and Theresa Magorian Friedlmeier. 2017. "Advances in Cost-Efficient Thin-Film Photovoltaics Based on Cu(In,Ga)Se2." *Engineering* 3 (4): 445–451. https://doi.org/10.1016/J.ENG.2017.04.015.

PV Education. 2019. "Absorption Coefficient | PVEducation." 2019. https://www.pveducation.org/pvcdrom/pn-junctions/absorption-coefficient.

Ratnaike, R. N. 2003. "Acute and Chronic Arsenic Toxicity." *Postgraduate Medical Journal* 79 (933): 391–396. https://doi.org/10.1136/pmj.79.933.391.

Rix, A. J., J. D. T. Steyl, J. Rudman, U. Terblanche, and J. L. van Niekerk. 2015. "First Solar's CdTe Module Technology – Performance, Life Cycle, Health and Safety Impact Assessment." Center for Renewable and Sustainable Energy Studies. http://www.firstsolar.com/-/media/First-Solar/Sustainability-Documents/Sustainability-Peer-Reviews/CRSES2015_06_First-Solar-CdTe-Module-Technology-Review-FINAL.ashx.

Solar Central. n.d. "Solar Markets Around The World." Accessed May 26, 2019. http://solarcellcentral.com/markets_page.html.

Wilkins, Tanya. 2018. "Alta Devices Sets Solar World Record - NASA Selects Alta Devices." *Alta Devices* (blog). December 12, 2018. https://www.altadevices.com/solar-world-record-nasa-selects-alta-devices/.

Xiao, Jianling, Hanlin Fang, Rongbin Su, Kezheng Li, Jindong Song, Thomas F. Krauss, Juntao Li, and Emiliano R. Martins. 2018. "Paths to Light Trapping in Thin Film GaAs Solar Cells." *Optics Express* 26 (6): A341–A351.

5

Third-Generation Solar Cells

Both first-generation photovoltaic (PV) technologies that embraced silicon and second-generation technologies that embraced semiconductor-based thin films are limited in energy conversion efficiency by the Shockley-Queisser limit (SQL) of 33.7% (Shockley and Queisser 1961). However, a new generation of PV technologies looks to surpass this limit by using novel device structures, new materials, concentrating optics, or a combination thereof. Such alternative materials have shown the potential to surpass existing commercially available PV products in terms of efficiency, cost, and robustness. Among the most promising of these alternative materials are organic, dye-sensitized, and perovskite materials, while Quantum Dot (QD) PV cells offer greater efficiencies through the use of extremely small, light-sensitive structures.

5.1 Organic PV Cells

In chemistry terminology, organic materials are those that contain carbon. Organic materials can be either natural or engineered and many are conjugated. A conjugated material is one in which the bonds that hold atoms together alternate between single and double bonds. This alternating bond structure allows atomic orbitals to overlap between adjacent atoms or molecules, thereby allowing electrons to be shared between atoms. The end result of these shared orbitals is a lower-energy and more stable system compared to similar materials that are not conjugated.

Because of the shared orbital structure inherent to conjugated organic PV materials, electrons are not associated with a single molecule or atom but instead with a localized system of multiple molecular orbitals. In contrast, electrons in semiconductors are associated with the valence band of only a single atom or molecule. When a semiconductor is exposed to light, these electrons are stimulated into a higher energy state within a range of energies called the conduction band. In contrast, when a conjugated organic material is exposed to light, electrons excited into higher energy states remain bound to localized states. These electrons are not yet free to travel through the material as is the case with electrons in the conduction band of a semiconductor. Further energy must be injected into the conjugated system to break an electron away from its localized state and enable it to travel or hop from one localized state to the next. Breaking up a photogenerated exciton into a pair of freely mobile charges (a hole and an electron) requires energy on the order of 100 meV (0.1 eV), which is significant compared to the few milli-electron volts that are required to break up electron-hole pairs in a semiconductor (Nelson 2003). The presence of these localized states and the additional energy required to enable electrons to hop between them is the fundamental difference between first- and second-generation PV cells and OPVs. For an electron in an organic material to be collected to produce useful electricity, a photogenerated electron must first dissociate from the hole in the exciton and then hop from one localized state to another until it reaches a contact and flows into an external circuit. One can think of a first- or second-generation PV cell as a highway, where electrons are unencumbered by traffic lights and with sufficient fuel (i.e., the force created by an electric field), travel easily over long distances. Organic PV (OPV) materials, on the other hand, are more like surface streets, where electrons and holes travel or hop from one traffic light and intersection to another. Hampered by traffic and other restrictions, the excitons in organic materials are less mobile than corresponding electron-hole pairs in conventional, semiconductor-based PV cells. Low mobilities combined with the extra energy required to break the electrostatic bond between the electron and hole in an exciton ultimately lead to dramatically reduced efficiencies for organic PV cells. In fact, the best performance of research cells to date have demonstrated maximum efficiencies of 15% for OPVs compared to the 26% efficiency demonstrated for comparable, single-junction monocrystalline silicon PV cells (National Renewable Energy Laboratory [NREL] 2019).

The energy band structure of an organic material is not described in terms of conduction and valence band energies. Instead, the band structure is characterized by two key molecular orbital energies. The first energy level of interest is that associated with the HOMO which is the highest occupied molecular orbital in dark (zero light) conditions. The LUMO energy, on the other hand, corresponds to the lowest unoccupied molecular orbital under zero light conditions. The difference between the LUMO and the HOMO energy levels is roughly equivalent to the bandgap of a semiconductor and like a semiconductor, determines the lowest energy photon (longest wavelength) of light that can be absorbed by the organic material.

Like a semiconductor, an organic PV device can be made with a single layer of organic material. The single layer consists simply of an organic PV material sandwiched between two electrical contacts. The contacts are chosen so that the difference in work functions between the two contacts sets up an electric field across the organic material and any free electrons are attracted to the positively charged contact while any free holes are attracted to the negatively charged contact. Just like a single-layer semiconductor, however, this electric field is weak and most electrons and holes recombine before reaching the contacts. Complicating matters further, electrons excited

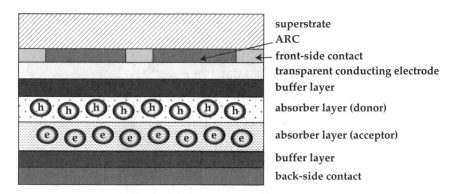

superstrate
ARC
front-side contact
transparent conducting electrode
buffer layer
absorber layer (donor)
absorber layer (acceptor)
buffer layer
back-side contact

FIGURE 5.1
Organic photovoltaic (OPV) cell structure. The PV cell structure consists of the active absorber layers (acceptor and donor), a buffer layer on front and back-side of the active layers, contacts on both sides of the cell, and a transparent conducting electrode. The transparent conducting electrode acts as a low-resistance interface between the active layers and the front-side contact. The buffer layer near the top of the cell protects the underlying absorber layers and facilitates hole transport to the front-side contact while blocking electron transport. The buffer layer near the back of the cell does the opposite, blocking hole transport to the back-side contact while facilitating electron transport.

by the incoming light remain bound to their respective holes and localized bonds restrict them from travelling freely through the organic PV material. As a result, the only excitons (i.e., bound electron-hole pairs) that make it to the contacts are those that diffuse there within 1–10 nm of the contact and even these may or may not be broken up or dissociated by the electric field. The end result is that devices based on single layers of organic materials are highly inefficient, resulting in quantum efficiencies of less than 1% and over-all energy conversion efficiencies less than 0.1% (Nelson 2003).

To improve efficiencies, organic materials, like inorganic semiconductors, can also be combined into bilayer (junction-based) devices for greater effi-ciency. An example of a single-junction, bilayer organic PV cell is shown in Figure 5.1. The organic PV cell relies on an absorber layer that consists of an acceptor material (comparable to an n-type semiconductor in a p-n junc-tion) and a donor material (comparable to a p-type semiconductor in a p-n junction).

In terms of energy (Figure 5.2), the LUMO and HOMO energy levels of the acceptor layers are lower than the LUMO and HOMO energy levels of the donor layers. In an organic material, this difference in LUMO and HOMO levels between acceptor and donor layers creates an electric field at the junc-tion between the two thin layers. The electric field has sufficient force to break apart photogenerated excitons and create free electrons and holes that can then travel to their respective contacts by floating up the energy hill (i.e., holes to the anode) or floating down the energy hill (i.e., electrons to the cathode).

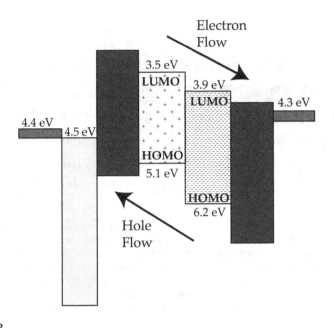

FIGURE 5.2
Organic photovoltaic (OPV) cell energy band structure. The energy band structure corresponds to the physical structure shown in Figure 5.1. In energy terms, holes float up the energy hill to the front-side of the cell (to the left in the above diagram) and electrons flow down the energy hill to the back of the cell (to the right). The LUMO and HOMO energy levels are roughly analogous to conduction band and valence band energy levels in a semiconductor. The numbers shown are the energy difference between the labelled energy level and the vacuum level and represent the work function of each layer. By design, any holes that attempt to float to the right are blocked by the buffer layer at the back of the cell and any electrons that try to flow down the energy hill to the front-side contact are blocked by the buffer layer near the top of the cell.

The number of excitons that break apart is dependent on how many can diffuse to the interface between donor and acceptor layers. Any excitons that do not reach the interface recombine and are not collected. The short diffusion lengths of these excitons has limited the efficiencies of PV cells. Best research cell efficiencies as of 2019 reached a maximum of 15.6% (NREL 2019).

An example of the layers that can be used to form an effective band structure for a single-junction, bilayer organic PV cell structure (Figure 5.1) is summarized in Table 5.1. The top layer (superstrate) of the PV cell is glass but can be replaced with plastic or other equally transparent material that enables highly flexible solar panels and structures that are well suited to wearable solar cell systems. Underneath the superstrate is a thin layer of indium tin oxide (ITO) that collects holes generated by light irradiating the underlying absorber layer. ITO is a very popular transparent conducting oxide (TCO) used as the front-side contact of PV cells to collect charge carriers. Alternatives to ITO include other TCOs like tin oxide and zinc oxide and

TABLE 5.1

Typical Composition of an Organic Photovoltaic (OPV) Cell

Layer	Material	Function
Superstrate	Glass	Passivates, protects, and provides structural stability to the PV cell
Front-side contact	Gold (Au)	Collects holes
Antireflective coating (ARC)	Moth-eye coating	Minimizes reflection of light away from the PV cell
Transparent conducting electrode	Indium tin oxide (ITO)	Provides a low-resistance contact to adjacent layers
Buffer layer	PEDOT-PSS (poly(3,4-ethylenedioxythiophene)/poly(styrenesulfonate))	Protects the active (absorber) layer; facilitates hole transport; blocks electron transport;
Active layer	Conducting or semiconducting polymer donor (e.g., poly(3-hexylthiophene)—P3HT)	Absorbs light; generates excitons
	Fullerene acceptor (e.g., Phenyl-C61-butyric acid methyl ester—PCBM)	
Buffer layer	Bathocuproine (BCP)	Facilitates electron transport; blocks hole transport
Back-side contact	Silver (Ag), aluminum (Al), magnesium (Mg)	Collects electrons

Source: Spooner (n.d.) and Forberich et al. (2008).

other materials including conductive polymers, very thin metal layers, and nanowires.

Underneath the ITO contact is a buffer layer designed to ensure an ohmic contact and effective hole transport from the underlying donor layer. In the context of PV cell performance, an ohmic contact is one in which current passes through the contact with very little voltage drop; in effect, the ohmic contract does not degrade or limit the performance of the PV cell. In contrast, other types of contact require a larger voltage drop to facilitate the flow of current, and for this reason impede PV cell performance. PEDOT-PSS (poly(3,4-ethylenedioxythiophene)/poly(styrenesulfonate)) is a common buffer layer used adjacent to a donor layer to improve the ohmic quality of the interface between the TCO and the rest of the PV cell and to support the transport of holes to the ITO contact. Buffer layers such as PEDOT-PSS also prevent oxygen from infiltrating the absorber layer and thereby protect the integrity and preserve the performance of the PV cell (Benanti and Venkataraman 2006).

Between the two buffer layers are the critical absorber layers responsible for absorbing light, creating excitons, and facilitating transport of electrons and holes to their respective contacts. The fullerene C60 and derivatives are common acceptor layers used in organic PV cells, while donor layers often consist of poly(phenylene vinylene) derivatives and poly(alkylthiophenes) (Benanti and Venkataraman 2006). The donor layer transports holes and the acceptor layer transports electrons to their respective contacts. The efficiency of the bilayer heterojunction structure can be enhanced by fabricating many (dispersed) junctions within a cell rather than fabricating a single planar junction.

5.2 Dye-Sensitized PV Cells

Unlike most PV technologies, the dye-sensitized solar cell (DSC) relies on a liquid (or in limited cases a solid) electrolyte in addition to diffusion processes to optimize PV behavior. The DSC consists of six basic layers, as shown in Figure 5.3: a superstrate (on top of the cell), a transparent conductive electrode that collects electrons, an electron extraction layer, a dye-sensitized layer that actively absorbs light, an electrolyte, and a counter electrode.

The dye-sensitized layer consists of a layer of an electron-transporting semiconductor and a separate layer of a photosensitive dye or, as shown in Figure 5.3, a porous electron-transporting semiconductor soaked in a photosensitive dye. The porous semiconductor approach provides a large surface area for the photosensitive dye to occupy and absorb incoming light.

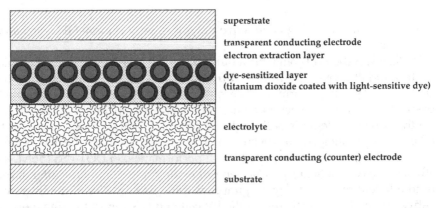

superstrate

transparent conducting electrode
electron extraction layer

dye-sensitized layer
(titanium dioxide coated with light-sensitive dye)

electrolyte

transparent conducting (counter) electrode

substrate

FIGURE 5.3
Dye-sensitized PV cell structure. The active layer of the dye-sensitized cell (DSC) structure consists of titanium dioxide particles coated with a photosensitive dye. When the dye absorbs light, the entire dye molecule (rather than a single electrons) moves to a higher energy level. Higher energy dye molecules readily donate electrons to the titanium dioxide particles and the electron extraction layer increases the number of electrons that are collected at the front-side of the cell (transparent conducting electrode). The counter electrode and the electrolyte replenish the electrons extracted by the titanium dioxide from the dye. The DSC is different from other PV cells in that it transports only free electrons and no holes.

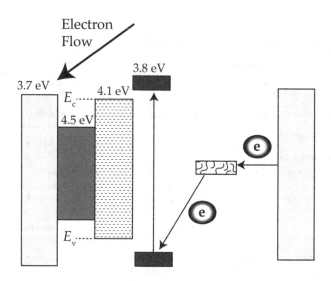

FIGURE 5.4

Dye-sensitized PV cell energy band structure. The energy band structure corresponds to the physical structure shown in Figure 5.3. When dye molecules absorb energy and move to a higher energy level, electrons in the dye molecules are then free to flow down the energy hill to the front-side of the cell. No free holes are generated in the DSC. The numbers shown are the energy difference between the labelled energy level and the vacuum level and represent the work function of each layer. The counter electrode and the liquid electrolyte in the back of the cell replenish electrons lost by the dye when absorbing photons of light.

Underneath the dye-sensitized layer is an electrolyte in contact with a counter electrode that completes the electrochemical cell contained within this PV cell architecture.

The energy band diagram of the DSC (Figure 5.4) is similar to other PV cells with three major exceptions. First, a photon of incoming light causes an entire dye molecule (as opposed to a single electron) to move from a lower energy to a higher energy level. The electron-transport layer (e.g., a porous semiconductor) easily extracts an electron from the energized dye molecule and delivers it to the front-side contact. Second, unlike many PV cell structures, a liquid electrolyte replenishes electrons in the dye molecules that were excited by light and transferred out of the PV cell through the front-side. And, finally, the DSC invokes only the flow of electrons (i.e., no holes) in the production of current.

The composition of a typical DSC is described in Table 5.2. Titanium oxide makes an inexpensive, readily available, porous electron transporter and ruthenium-based materials are excellent candidates for dyes because of their photosensitivity. A liquid electrolyte based on iodide is a common choice to replenish electrons in the ruthenium-based dye molecules. And, the counter electrode may or may not be coated with a catalyst like platinum to promote the transfer of electrons to the electrolyte and subsequently, the entire PV cell. To ensure

TABLE 5.2

Typical Composition of a Dye-Sensitized Cell (DSC)

Layer	Material	Function
Superstrate	Glass	Provides structural stability; seals the cell
Electron extraction layer	Graphene, tungsten oxide	Increases collection of electrons
Transparent conducting electrode	Fluorine tin oxide (FTO)	Collects electrons
Dye-sensitized layer	Titanium oxide (TiO_2) nanoparticles soaked in a photosensitive ruthenium dye	Dye molecules absorb light; TiO_2 absorbs electrons from dye molecules and transports electrons
Electrolyte	Iodide (I^-); tri-iodide (I_3^-)	Replenishes lost electrons in dye molecules
Counter electrode	Various metals and alloys, carbon materials, conductive polymers, transition metal compounds	Completes the electrochemical cell

Source: Chen (2015) and Zheng, Tachibana, and Kalantar-Zadeh (2010).

that the liquid electrolyte does not leak out of the cell, the entire DSC is fully sealed after fabrication.

Because the spectral absorption of the photosensitive dyes used in DSCs is not as compatible with sunlight as silicon is, the efficiency of these devices for conventional solar energy applications can be limited. As of 2019, the best DSCs reached optimal efficiencies in response to sunlight (AMI 1.5) of about 24% (NREL 2019). However, this incompatibility with sunlight can be an advantage for wearable PV cells that rely on both artificial and natural (sunlight) sources to operate. Artificial light sources tend to have less red in their spectra than sunlight, thus making them more compatible with DSCs than silicon.

DSCs, like other thin-film technologies, are more mechanically robust than traditional silicon PV cells. Thus, they require less protection and the glass superstrate is often replaced with a thin layer of plastic in packaged panels. This thin overlayer allows far more heat to be released during operation, which in turn, minimizes efficiency losses caused by self-heating that are common in silicon solar panels.

Because DSCs only transport electrons and not holes, there is also one less opportunity for electrons to recombine which ultimately means that a higher percentage of excited electrons are collected as current. The absence of photogenerated holes is especially advantageous in the low-light or variable conditions that are consistent with artificial light sources and wearable PV applications. Under these low-light conditions, many PV technologies "cut out" when leakage and recombination overwhelm any photogenerated electrons and holes. DSC PV cells do not cut out and continue to operate under low-light conditions.

The main disadvantage of the DSC is the liquid electrolyte that requires the cell to be sealed during manufacturing and remain sealed over its lifetime. At too low a temperature, the electrolyte can freeze which drops power production to zero and may also physically damage the DSC panel. At high temperatures, the electrolyte can expand, generate cracks, and leak out of the solar panel altogether. Fortunately, in wearable solar cell applications, extreme temperatures are far less common than in conventional outdoor solar panel installations, which dramatically reduces the risk of temperature-induced failure. DSCs are also hampered by the high cost of platinum when it is used as a catalyst on the counter electrode. Higher costs limit the economic competitiveness of DSCs with other low-cost PV technologies. However, alternatives to platinum-coated electrodes are a subject of recent research aimed at more fully realizing the low-cost potential of the DSC (Iqbal and Khan 2018).

5.3 Perovskites

Another third-generation family of PV cell materials is the perovskites, named after the perovskite mineral structure. The perovskite mineral is arranged in a cube-like structure and contains calcium, titanium, and oxygen ($CaTiO_3$) arranged as one type of large positively charged ion in the center of the cubic structure, another type of positively charged ion located in the corners, and smaller, negatively charged ions of yet another type along the faces of the cube. In the world of PV technology, the term perovskite refers to any compound that has a similar structure to the mineral itself. And, by manipulating the three types of ions that make up the perovskite structure, a wide range of desirable properties can be obtained including PV sensitivity (Fan, Sun, and Wang 2015), superconductivity (He et al. 2001), and giant magnetoresistance (Shimakawa, Kubo, and Manako 1996). When developed for PV cells, the ion in the center of the perovskite cube is typically a large, organic ion (e.g., methylammonium or formamidinium), the ions on the corners are a large and inorganic molecule like lead, and the smaller ions on the faces of the cube are halogen ions such as chloride or iodide (Fan, Sun, and Wang 2015).

The basic structure of the perovskite solar cell (PSC) is shown in Figure 5.5. It is similar to the DSC structure except that no electrolyte is involved and the PSC transports both electrons (to the transparent electrode) and holes (to the back electrode). Transport layers facilitate both the movement of electrons and holes.

An energy band structure for a perovskite PV cell is shown in Figure 5.6 based on the layers described in detail in Table 5.3 and the structure of Figure 5.5. Electrons generated from the perovskite when light is

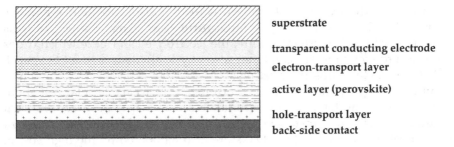

superstrate

transparent conducting electrode

electron-transport layer

active layer (perovskite)

hole-transport layer
back-side contact

FIGURE 5.5
Perovskite PV cell structure. The perovskite PV cell structure is similar to that of the amorphous silicon PV cell (Figure 3.4). Perovskites are strong light absorbers and they act similar to the insulator layer in a p-i-n junction by providing significant volume for light absorption and the subsequent generation of excitons. Excitons are broken apart into free electrons and free holes. The free holes travel through the hole transport layer to the bottom of the cell for collection while the free electrons travel through the electron layer so the top of the cell for collection. The planar structure shown above can be replaced with a mesoporous layer where perovskite coats titanium dioxide particles, similar to the dye-sensitized cells (DSC) structure shown in Figure 5.3.

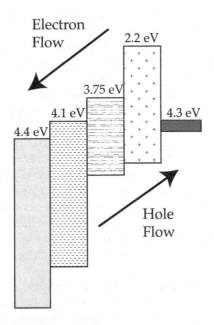

FIGURE 5.6
Perovskite PV cell energy band structure. Electrons flow down the energy hill from the active perovskite layer to the electron-transport layer and to the transparent conducting electrode for collection. Holes float upward in the opposite direction toward the back of the cell and are collected at the back-side contact.

TABLE 5.3

Typical Composition of a Perovskite Solar Cell (PSC)

Layer	Material	Function
Superstrate	Glass	Provides protection and structural stability
Transparent conducting electrode	Fluorine tin oxide (FTO)	Collects electrons
Electron-transport layer	Titanium dioxide (TiO2)	Transports electrons
Active layer	$CH_3NH_3PbI_{3-x}Cl_x$ perovskite	Absorbs light
Hole transport layer	Spiro-OmeTAD	Transports holes
Back electrode	Silver (Ag)	Collects holes

Source: Da, Xuan, and Li (2018).

absorbed flow down the energy hill to the left toward the transparent electrode with the assistance of a titanium dioxide electron-transport layer. On the other side of the PSC, a hole transport layer facilitates the floating of holes upward to the energy level of the silver back electrode.

In addition to strong absorption characteristics and high efficiencies, the three molecules that make up the PSC structure can be chosen to adjust the bandgap of the perovskite and thereby tune the absorption spectrum to better match the visible and near-infrared light spectrum of incoming light. PV cells made from perovskite materials can thus be matched to a variety of natural and artificial light sources for maximum performance in specific environments. In contrast, bandgap tuning is not possible with silicon and many other semiconductor PV technologies.

Rapid advances in deposition techniques and advances in perovskite materials have enabled their research cell efficiencies to increase dramatically in the past few years from 13% in 2013 to 24.2% in 2019 (NREL 2019). The high absorption coefficients of perovskites have enabled very thin PV cells, on the order of hundreds of nanometers, to be feasible and practical. Very thin films, alongside low processing and raw material costs, make possible a wide range of flexible, lightweight, and inexpensive wearable solar cell systems and footprints.

Particularly noteworthy for the production of wearable PV products is the low-temperature ink-jet printing of perovskite materials into PV structures onto thin foils (Saule Technologies n.d.). Ink-jet printing not only allows for quick prototyping and low-cost manufacturing but also enables control of film morphology and homogeneity through key printing parameters and adjustment of wetting characteristics and chemistry through precursor solution adjustments. Not only do ink-jet printing and other low-complexity fabrication methods decrease the manufacturing costs of PSCs, but they also support the rapid prototyping of wearable devices that use solar cells and systems to generate electrical energy and supply power.

Despite their tremendous advances in efficiency in recent years, the long-term stability of perovskites remains unpredictable, which limits their application in conventional or wearable PV devices that are expected to last for years. Advances in electron-transport layers have reduced degradation in the presence of sunlight (Hutchins 2019). But, many perovskites degrade dramatically in the presence of moisture and can also degrade internally when heated. Recently discovered hysteresis in perovskite PV cells is also of concern. Improvements in both stability and hysteresis are a topic of continuing research (Leijtens et al. 2015).

5.4 Quantum Dot PV Cells

A QD is a particle whose size is smaller than the exciton Bohr radius of the semiconductor from which it is fabricated. The exciton Bohr radius is defined as the most likely distance between an electron and a hole in an electron-hole pair that results from injection of energy (e.g., light) into the atom. The exciton Bohr radius is dependent on the effective mass of electrons and holes in the semiconductor, is on the order of nanometers, and is tabulated for common QDs in Table 5.4. As the size of the QD shrinks below the exciton Bohr radius, energy states become discrete rather than existing in continuous energy bands (e.g., conduction and valence bands) as is the case in a traditional semiconductor. Discrete energy states mean that electrons excited by light can move only to certain other energy levels rather than almost any energy level that is greater than the conduction band energy. Further, these discrete energy levels can be tuned by adjusting the size of the QD.

The smaller the QD, the more sensitive it is to shorter wavelengths. As the size of a QD decreases, the bandgap increases, and the maximum wavelength

TABLE 5.4

Exciton Bohr Radius and Bandgap for Common Quantum Dot Photovoltaic (QD PV) Cell Materials

Quantum Dot Semiconductor	Exciton Bohr Radius (nm)	Bandgap E_g (eV)		
		QD Radius >40 nm	QD Radius = 10 nm	QD Radius = 5 nm
PbS (lead sulfide)	40.0	0.41	0.6	1.0
GaAs (gallium arsenide)	28.0	1.43	1.8	3.0
CdTe (cadmium telluride)	15.0	1.50	1.9	2.7
CdSe (cadmium selenide)	10.6	1.74	2.0	2.8
CdS (cadmium sulfide)	5.6	2.53	2.6	3.0

Source: Jasim (2015).

of light that the QD can absorb decreases according to the Planck-Einstein relation:

$$\lambda_{max} = \frac{hc}{E_g} \tag{5.1}$$

where h is Planck's constant (4.14×10^{-15} eV sec), c is the speed of light (3×10^8 m/sec in air), and E_g is the bandgap in electron volts.

With this kind of tunability, QDs can be made into PV cells that absorb light from a large proportion of the sunlight spectrum, particularly ultraviolet (UV) and infrared wavelengths that are not absorbed efficiently by other types of PV materials. The tunable bandgap can also be used to make PV cells that are best matched to harvest light from other sources of light. This possibility is not particularly important for conventional solar panels that are almost without exception optimized to harvest natural sunlight for energy. However, for wearable solar cell systems, the QD PV cell can be optimized for the dominant light source in the user's environment whether that be natural sunlight or a particular artificial light source (e.g., fluorescent, incandescent).

Also, because of the discrete energy levels that are characteristic of QDs, it becomes possible for a single photon of light to generate more than one electron. This process, called multiple exciton generation (MEG) is a result of impact ionization by excited electrons. In most semiconductors, an electron excited into the conduction band by a photon of light relaxes to the bottom of the conduction band in processes that release heat. In a QD, however, the excited electron can collide with other electrons, imparting energy to those electrons that is sufficient for those electrons to be collected as current. This multiplicative effect caused by impact ionization and the resulting production of multiple excitons for a single photon makes it possible for QD PV cells to exceed the SQL of 33.7% efficiency (Shockley and Queisser 1961). Challenges in developing processes for the synthesis and production of QD films that reduce traps, defects, and other places where electrons can get lost before collection have held QDs back from surpassing the SQL. In 2019, the best research cell efficiency for a quantum dot (QD) was recorded at 16.6% (NREL 2019).

Because QDs are tiny, they are also space efficient. A QD layer in a PV cell can consist of as few as 100 atoms, potentially leading to very thin PV cells that are highly flexible, and with certain fabrication processes, very inexpensive. QDs can be fabricated into PV cells in a number of different ways, many of which mimic existing PV cell structures. For example, QDs can be inserted into DSC structures similar to Figure 5.3 using layers of material such as those described in Table 5.5. In place of a photosensitive dye molecule moving to a higher energy level upon absorption of light, the QD particle absorbs light, causing electrons to be excited to one of the discrete higher energy levels available in the QD. Once in this excited state, electrons can either impact ionize other electrons or flow into the titanium dioxide (electron-transport layer)

TABLE 5.5

A Typical QD-Sensitized PV Cell

Layer	Material	Function
Superstrate	Glass	Provides structural stability and seals the cell
Transparent conducting electrode	Fluorine tin oxide (FTO)	Collects electrons
Electron-transport layer	Titanium dioxide (TiO_2)	Transports electrons to the transparent electrode
Active layer	CdS, CdSe QDs	Absorbs light; generate excitons
Electrolyte	Polysulfide electrolyte	Replenishes electrons
Counter electrode	Brass coated with copper sulfide (Cu_2S)	Returns electrons to the electrochemical cell and catalyzes (promotes) the process; reflects light back into the PV cell

Source: Jasim (2015), Pan et al. (2018), and Wu et al. (2017).

and then into the transparent electrode at the top of the PV cell. Meanwhile, the electrolyte (liquid or solid) underneath the QD layer replenishes the electrons that the QDs have lost to the electron-transport layer. QD-sensitized PV cells using QDs alone and in conjunction with photosensitive dyes have been demonstrated in the literature (Jasim 2015).

As with DSCs, liquid electrolytes are difficult to manage and limit the reliability of resulting solar modules. Solid electrolytes are difficult to find. To address these issues, QD PV cells can also be made without an electrolyte. The resulting structure is similar to the PSC (Figure 5.5). A typical QD PV heterojunction (Table 5.6) made with this structure contains a QD film fabricated adjacent to a semiconducting oxide layer. Just as in the DSC, this layer serves as a means to transport electrons from the QD layer to the transparent electrode. On the other side of the QD layer, however, the electrolyte layer is eliminated and an optional hole transport layer facilitates the movement of holes generated in the QDs to the back-side contact in the PV cell.

TABLE 5.6

Example of a Quantum Dot (QD) Heterojunction PV Cell

Layer	Material	Function
Superstrate	Glass	Provides protection and structural stability
Transparent conducting electrode	Indium tin oxide	Collects electrons
Electron-transport layer	Zinc oxide (ZnO)	Transports electrons
Active layer	Lead sulfide (PbS) QDs	Absorbs light
Hole transport layer	Molybdenum oxide (MoO_3)	Transports holes
Back-side contact	Gold (Au)	Collects holes

Source: Sogabe, Shen, and Yamaguchi (2016).

Several other architectures for PV cells can support QD PV cells. Metal/semiconductor junctions, hybrid QD/silicon cells, and p-i-n structures (similar to amorphous silicon PV cells) are available and their development is rapidly evolving to take advantage of advanced structures and fabrication/assembly methods (Jasim 2015). However, the degradation of QD solar cells in the presence of moisture and UV light continues to limit their commercial viability and useful lifetime. And, the inherent toxicity of the most common QDs, containing such heavy metals as cadmium, selenium, or lead, make them highly toxic and environmentally harmful at the end of their life cycle. Their toxicity limits their appeal for wearable PVs.

5.5 Summary

In different ways, the third-generation technologies described in this chapter are capable of exceeding the SQL of 33.7% established for first- and second-generation PV cells based on inorganic semiconductors. In QD PV cells, for example, a single photon can generate multiple current carriers via MEG processes. As a result, the ultimate efficiency of QD-based single-junction cells approaches 44% (Sogabe, Shen, and Yamaguchi 2016). Among organic PV cells, the potential to exceed the SQL is still a matter of debate. Conventional wisdom dictates that the single-junction, organic PV cells should possess ultimate efficiencies that are lower than the SQL because of the additional energy required to split apart an electron and a hole in a photogenerated exciton. However, the recent literature has challenged this assumption by exploring singlet exciton fission (where one photon can produce two pairs of electrons and holes in organic PV cells) and field-dependent recombination. In the latter, maintaining a high electric field at the junction between donor and acceptor layers (Figure 5.1) reduces the recombination of electrons and holes and increases the number of current carriers that can be collected at the contacts, thereby increasing the fill factor of the cells and increasing the maximum possible efficiency (Trukhanov, Bruevich, and Paraschuk 2015). In perovskite-based technologies, hot carrier collection is the primary route toward exceeding the SQL. Recent advances in perovskite PV cells have demonstrated the potential to harvest more current by extending the cooling time and the transport time of hot carriers to tens of picoseconds and hundreds of nanometers, respectively. When hot carriers take longer to thermalize or cool back to lower-energy states, they can travel farther and, in so doing, are more likely to reach PV cell contacts where they can be harvested and collected as current (Guo et al. 2017). Other PV cell technologies, including DSCs, can achieve efficiencies beyond the SQL through device structures that include multiple junctions, tunable bandgap materials, alternative optics designs or a combination thereof. In total, these third-generation PV technologies behave very differently from first- and second-generation technologies, promising efficiencies that have yet to be fully exploited.

References

Benanti, Travis L., and D. Venkataraman. 2006. "Organic Solar Cells: An Overview Focusing on Active Layer Morphology." *Photosynthesis Research* 87 (1): 73–81.

Chen, Lung-Chien. 2015. "Dye-Sensitized Solar Cells with Graphene Electron Extraction Layer." In *Optoelectronics - Materials and Devices*, edited by Sergei L. Pyshkin and John Ballato. London, England: IntechOpen Limited. https://doi. org/10.5772/60644.

Da, Yun, Yimin Xuan, and Qiang Li. 2018. "Quantifying Energy Losses in Planar Perovskite Solar Cells." *Solar Energy Materials and Solar Cells* 174: 206–213.

Fan, Zhen, Kuan Sun, and John Wang. 2015. "Perovskites for Photovoltaics: A Combined Review of Organic–Inorganic Halide Perovskites and Ferroelectric Oxide Perovskites." *Journal of Materials Chemistry A* 3 (37): 18809–18828. https://doi.org/10.1039/C5TA04235F.

Forberich, Karen, Gilles Dennler, Markus C. Scharber, Kurt Hingerl, Thomas Fromherz, and Christoph J. Brabec. 2008. "Performance Improvement of Organic Solar Cells with Moth Eye Anti-Reflection Coating." *Thin Solid Films*, Proceedings on Advanced Materials and Concepts for Photovoltaics EMRS 2007 Conference, Strasbourg, France, 516 (20): 7167–7170. https://doi.org/10.1016/j.tsf.2007.12.088.

Guo, Zhi, Yan Wan, Mengjin Yang, Jordan Snaider, Kai Zhu, and Libai Huang. 2017. "Long-Range Hot-Carrier Transport in Hybrid Perovskites Visualized by Ultrafast Microscopy." *Science* 356 (6333): 59–62. https://doi.org/10.1126/science.aam7744.

He, T., Q. Huang, A. P. Ramirez, Y. Wang, K. A. Regan, N. Rogado, M. A. Hayward, et al. 2001. "Superconductivity in the Non-Oxide Perovskite MgCNi 3." *Nature* 411 (6833): 54. https://doi.org/10.1038/35075014.

Hutchins, Mark. 2019. "It's All in the Electrons – Japanese Scientists Give Perovskites a Push." *PV Magazine International*, January 23, 2019. https://www.pv-magazine.com/2019/01/23/its-all-in-the-electrons-japanese-scientists-give-perovskites-a-push/.

Iqbal, Muhammad Zahir, and Sana Khan. 2018. "Progress in the Performance of Dye Sensitized Solar Cells by Incorporating Cost Effective Counter Electrodes." *Solar Energy* 160 (January): 130–52. https://doi.org/10.1016/j.solener.2017.11.060.

Jasim, Khalil Ebrahim. 2015. "Quantum Dots Solar Cells." In *Solar Cells - New Approaches and Reviews*. London, England: IntechOpen Limited. https://doi.org/10.5772/59159.

Leijtens, Tomas, Giles E. Eperon, Nakita K. Noel, Severin N. Habisreutinger, Annamaria Petrozza, and Henry J. Snaith. 2015. "Stability of Metal Halide Perovskite Solar Cells." *Advanced Energy Materials* 5 (20): 1500963.

Nelson, Jenny. 2003. "Organic and Plastic Solar Cells." In *Practical Handbook of Photovoltaics: Fundamentals and Applications*, 484–511. Boca Raton, Florida: CRC Press.

National Renewable Energy Laboratory (NREL). 2019. "Best Research-Cell Efficiency." https://www.nrel.gov/pv/cell-efficiency.html.

Pan, Zhenxiao, Huashang Rao, Iván Mora-Seró, Juan Bisquert, and Xinhua Zhong. 2018. "Quantum Dot-Sensitized Solar Cells." *Chemical Society Reviews* 47 (20): 7659–7702. https://doi.org/10.1039/C8CS00431E.

Saule Technologies. n.d. Solar Cells Reimagined. Accessed May 12, 2019. https://sauletech.com/.

Shimakawa, Y., Y. Kubo, and T. Manako. 1996. "Giant Magnetoresistance in Ti2Mn2O7 with the Pyrochlore Structure." *Nature* 379 (6560): 53. https://doi.org/10.1038/379053a0.

Shockley, William, and Hans J. Queisser. 1961. "Detailed Balance Limit of Efficiency of pn Junction Solar Cells." *Journal of Applied Physics* 32 (3): 510–519.

Sogabe, Tomah, Qing Shen, and Koichi Yamaguchi. 2016. "Recent Progress on Quantum Dot Solar Cells: A Review." *Journal of Photonics for Energy* 6 (4): 040901. https://doi.org/10.1117/1.JPE.6.040901.

Spooner, Emma. n.d. "Organic Photovoltaics: An Introduction." Ossila. Accessed May 24, 2019. https://www.ossila.com/pages/organic-photovoltaics-introduction.

Trukhanov, Vasily A., Vladimir V. Bruevich, and Dmitry Yu Paraschuk. 2015. "Fill Factor in Organic Solar Cells Can Exceed the Shockley-Queisser Limit." *Scientific Reports* 5 (June): 11478. https://doi.org/10.1038/srep11478.

Wu, Jihuai, Zhang Lan, Jianming Lin, Miaoliang Huang, Yunfang Huang, Leqing Fan, Genggeng Luo, Yu Lin, Yimin Xie, and Yuelin Wei. 2017. "Counter Electrodes in Dye-Sensitized Solar Cells." *Chemical Society Reviews* 46 (19): 5975–6023. https://doi.org/10.1039/C6CS00752J.

Zheng, Haidong, Yasuhiro Tachibana, and Kourosh Kalantar-Zadeh. 2010. "Dye-Sensitized Solar Cells Based on WO3." *Langmuir: The ACS Journal of Surfaces and Colloids* 26 (24): 19148–19152. https://doi.org/10.1021/la103692y.

6

Arrays of PV Cells

A direct connection between a photovoltaic (PV) cell and a battery, a portable, or a wearable invites trouble. The voltage produced by a single PV cell is so low that most batteries will not notice the cell is there, much less connected and supplying power. Furthermore, the current produced by a single cell is sufficient to power only the smallest of wearables. And, the fluctuations in current with the lighting conditions inherent to wearable systems would drive a battery to a certain and early death. For these reasons, PV cells in wearable systems must turn to arrays of PV cells to supply a voltage and current that is compatible with the devices they strive to power. Connected in series, strings of PV cells produce higher voltages, and connected in parallel, multiple strings of PV cells produce higher current and greater power.

Placed in an array and working as a team, PV cells tirelessly and synergistically transform the energy contained in light into useful electricity. Unfortunately, such synergy only fully materializes when the PV cells are both identical to one another and working under identical irradiance (light intensity). Too often, however, when the real world intrudes on this happy display of PV teamwork, individual cells experience manufacturing mismatch and different levels of shading, soiling, aging, and a whole host of other variables that require electronics to come to the rescue. Support and control electronics manage an array of PV cells by tracking maximum power conditions, by avoiding hot spots that can cause fire, damage, and drama, and in some cases, by reconfiguring arrays on the fly to better match neighboring cells for greater power generation. Together, well-designed PV arrays and support electronics produce substantial and consistent power and experience fewer catastrophic failures among PV cells in all kinds of solar energy systems.

6.1 Basic PV Array Design

One of the first decisions that goes into designing a solar panel is determining how many PV cells a panel will contain and how the cells will be configured in the panel. A typical solar panel consists of multiple strings, each containing a fixed number of series-connected PV cells. These strings are then connected in parallel with one another to complete the array.

Strings of PV cells increase the array output voltage to a level compatible with energy storage devices (e.g., batteries). With the exception of gallium arsenide (GaAs) and perovskite cells, most PV cells operate at a maximum voltage between 0.5 and 0.7 V. In theory, it is possible to use boost converter electronics to increase the voltage of a single PV cell to a level sufficient to support energy storage or directly power some electronic devices. Unfortunately, the efficiency of this approach is often poor and results in significant power losses. Further, these power losses only increase with an increasing gap between PV cell voltage and device/battery voltage. Thus, the losses involved in powering the higher supply voltages required of more power-hungry devices like laptop computers would be much greater than those associated with powering smaller, less power hungry wearables. Addressing this problem and limiting power losses so that they are independent of supply voltage requires configuring arrays of PV cells to provide a voltage close to what batteries need to charge and what electronic devices need to function properly.

To resolve inefficiencies associated with conditioning the voltage of a PV cell array, PV cells are connected in series to form strings. Strings produce a voltage that is the sum of the cell voltages in the string (Figure 6.1). The current flowing through a PV string must be the same for all the cells in the string. For PV cells with similar characteristics operating under similar lighting conditions, series connections result in little loss of power because

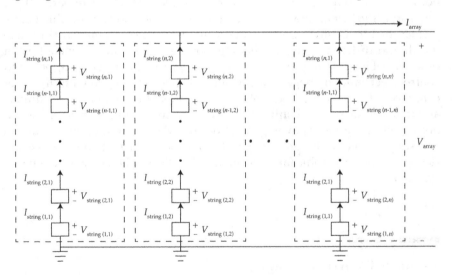

FIGURE 6.1
PV cell arrays. In most solar panels, PV cells are first connected in series and the resulting strings are connected in parallel to form PV arrays (the array shown here has equal numbers of cells in series and parallel, but this is not necessarily the case in most PV panels). The currents flowing through PV cells in the same string are always equal, but the voltage across the string is the sum of the voltages across individual PV cells within the string. The array current is the sum of all the string currents and the array voltage is an average of all the string voltages.

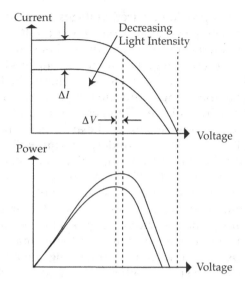

FIGURE 6.2
PV cell characteristics under changing irradiance conditions. As the amount of light energy reaching a PV cell decreases, less power is produced. At the MPP, the cell conducts less current and has less voltage across it. However, the reduction in current is significantly greater than the reduction in voltage that arises with the shift in operating point.

the operating characteristics and maximum power point (MPP) of all of the cells in a string are similar.

Fortunately, PV cell voltage does not change much as irradiance (light intensity) and input conditions change. As a result, the voltage produced by an entire string fluctuates little if all cells continue to operate at or near their MPP (Figure 6.2). In contrast, current fluctuates much more widely as a result of mismatches in PV cell characteristics, differences in the irradiance on each cell, and shading, soiling, and other irregularities associated with individual cells. Such variations can drastically reduce the current flowing through the string because string current is always limited by the weakest cell in the string. Furthermore, as healthy PV cells adapt to the smaller current and reduced power of a weak or underperforming cell in a string, they can cause the under-performing cell to develop a negative voltage (i.e., reverse bias). In this situation, the underperforming cell starts to consume power rather than producing power, which compounds power losses resulting from reduced currents already flowing through the string. Reverse biased, underperforming cells are also vulnerable to overheating which causes temporary and sometimes permanent damage in addition to posing a fire and safety hazard. Avoiding these "hot spots" and other consequences of underperforming cells in a string requires that PV strings be managed using control electronics specifically designed to meet the needs of the string when it is exposed to shading, soiling, and other ambient variations in cell behavior during everyday operations.

Strings of PV cells provide desired voltages for subsequent electronic devices or energy storage systems. Connecting the PV strings in parallel complements the function of a single string by boosting current produced from the PV array (Figure 6.1). The total current produced by multiple strings in parallel is simply the sum of the individual string currents, while the voltage of the strings connected in parallel is approximately the average of all the string voltages. The penalty for a PV string producing a voltage dissimilar to other strings in the array is not nearly as severe as the current penalty imposed by an underperforming PV cell in a string for three reasons: (a) PV cell and PV string voltages do not vary as much as PV cell currents (Figure 6.2); (b) PV array voltage approaches the average string voltage rather than the worst case string voltage as is the case with cell currents within a string; and (c) most DC-DC converters can handle small drops in input voltage without significant decreases in efficiency. For these reasons, electronics designed to better manage the operation of PV arrays focus their attention on mitigating the effects of underperforming currents within strings rather than on underperforming voltages across strings.

Traditional solar cell systems installed on rooftops or other stationary locations use solar panels with thirty-six 15.2 cm × 15.2 cm PV cells connected in series to support a nominal 12-V output and seventy-two 15.2 cm × 15.2 cm PV cells in series to support a nominal 24-V output (Samlex Solar n.d.). Based on a typical operating voltage of 0.5 V per cell, 36 cells provide 18 V that can be downregulated to the charging profiles required of a standard 6-cell, 12-V lead-acid battery. Similarly, 72 PV cells in series can be downregulated to support 24-V energy storage. The reason that PV cell arrays are designed to produce more voltage than what is needed for energy storage is to account for deterioration in output voltage caused by self-heating, partial shading of PV cells, and other environmental effects (PV Education 2019).

Portable solar chargers require fewer PV cells in series because of the lower voltages they support—around 5 V. Smaller strings, consisting of eight smaller PV cells connected in series, are more than adequate to support the 3.7 V–4.2 V required to charge a single Li-ion battery (Falin and Li 2011). An even smaller number of cells can be used when a boost (step-up) converter is used to condition the PV array voltage to the voltage or current needed for optimal battery charging. Smaller strings have a number of advantages. First, a single underperforming cell compromises the performance of a much smaller number of cells (per string) resulting in lower overall power losses. And, since PV cells in a string occupy a much smaller area in a smaller string, it is much less likely that one cell will be significantly more shaded, soiled, or otherwise more compromised than other cells in the same string. In wearable solar cell systems, where solar panels are much smaller than traditional systems with correspondingly smaller individual PV cell area, the likelihood of extreme underperformers caused by disparities in shading, soiling, or other irregularities within a string drops even more. Nevertheless, in solar cell systems that use flexible PV cells, wearable or otherwise, bending and other movement in individual PV cells can result in different orientations of cells within a string that produce similar disparities in performance as partial shading and soiling.

6.2 Array Management

Many PV arrays are overdesigned to produce a greater nominal voltage than what subsequent energy storage devices need to operate properly. These higher voltages can always be stepped down to the exact voltage required by a battery or other energy storage device through the use of a buck converter. However, it is also possible to step up the output voltage of a PV array using a boost converter that has the added benefit of requiring fewer PV cells in series. Buck and boost converters are discussed further in Chapter 7.

PV arrays are designed to provide input to either a boost or buck converter (but usually not both). While both types of converters allow some flexibility in how high or how low the PV array voltage can go, array management must still mitigate the effects of underperforming cells on the overall power production. Blocking and bypass diodes are a key part of the basic strategy to managing underperforming cells. A bypass diode (Figure 6.3a) is used to prevent an underperforming cell from sinking power rather than producing it and becoming a hot spot that is vulnerable to cracking and other physical damage. The bypass diode is connected parallel to the PV cell so that if the string causes the PV cell to go into reverse bias, the diode goes into forward bias and conducts the string current that might have otherwise flowed through the underperforming PV cell and damaged it. In large-scale solar cell systems where strings conduct appreciable amounts of current, diodes are often too expensive to bypass single cells, so they are typically connected across multiple cells in a string. If a cell in the substring underperforms, the bypass diode shuts down the power production in all of the cells in the substring. In wearable solar cell systems, the current generated is smaller, the required diodes are smaller and less expensive, and single cell bypass circuits are affordable and effective at reducing the power loss incurred by the cells that underperform.

While bypass diodes are used to manage PV cells at a cell or substring level, other diodes, called blocking diodes, operate at the whole string level (Figure 6.3). While bypass diodes are connected in parallel with a single or small substring of PV cells, blocking diodes are connected in series with an entire string of cells. If for any reason, a particular string loads down the array and attempts to sink power rather than produce it, the blocking diode turns off and prevents current from flowing backward into the string. While both blocking and bypass diodes are effective at preventing excessive power

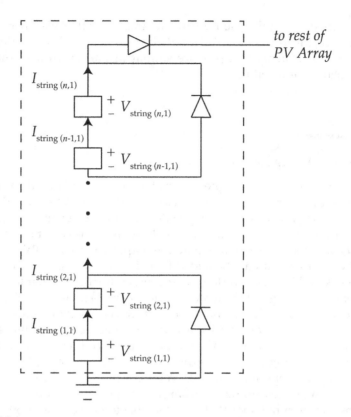

FIGURE 6.3
Blocking and bypass diodes in PV cell arrays. Bypass diodes operate in parallel to a single PV cell or small number of PV cells (shown as a substring of two cells in the above figure). If a PV cell is underperforming and generating a weak current, the bypass diode will turn on to prevent the cell from going into reverse bias and overheating. Blocking diodes operate in series with each PV string in an array and prevent current from flowing backward into the string.

drain or damage from underperforming cells, they are still passive elements and can consume appreciable power as they do so.

Alternatives to blocking and bypass diodes have been developed to overcome the disadvantages of these basic approaches. One viable alternative to the use of multiple diodes in a PV array for wearable solar cell systems is to prefer parallel over series connections of PV cells. Parallel connections address the nonuniformities in PV cell behavior and are easier to match when the behavior of PV cells varies frequently and broadly. When a PV cell array is dominated by parallel rather than series connections, current-balancing problems are minimized. The reduced output voltage of parallel connections can cause a boost converter to have lower overall efficiency, but these reductions in efficiency may be tolerable given the consequences of underperformers in arrays containing larger strings with more PV cells connected in series per string.

6.3 Maximum Power Point Tracking

No matter the size of the array or the conditions under which it operates, well-managed PV arrays include circuits and algorithms that find the operating point corresponding to maximum or near-maximum power. True maximum power is only possible when PV cells respond identically to one another and are exposed to identical irradiance (light intensity) conditions. Realistically, only near-maximum power is within reach by using maximum power point tracking (MPPT). MPPT searches for and locates the MPP for each PV cell, each string of cells, or each predetermined subset of cells in a PV array. The goal of MPPT is to maximize the power production of the array given the limitations of PV cells operating within it. A wide variety of MPPT techniques are available, most of them developed for grid-connected solar cell systems (Esram and Chapman 2007). Hill climbing techniques are among the most popular approaches to MPPT. In the perturb and observe (P&O) approach to hill climbing, the operating voltage or current of a PV cell or a PV array is increased or decreased and the product of voltage and current (i.e., the power) computed. If the power is more than that produced at the original operating voltage and current before the perturbation, the voltage or current is increased once again. If the power is less, the voltage or current is then adjusted in the opposite direction. The efficiency of this approach in finding the maximum power is over 96% (Rawat and Chandel 2013). To accommodate rapidly changing lighting conditions that often occur on partly cloudy days or are the rule rather than the exception for wearable solar cell systems, adaptive P&O may be better suited to MPPT. In this modified P&O approach, adjustments are larger when irradiance is changing quickly in order to speed up the rate at which a PV cell or array finds the MPPT. When conditions are more stable, perturbation changes are kept smaller in order to avoid overshooting the MPP and wasting power and time in the process (Rawat and Chandel 2013). Neither of these P&O approaches allows the MPP to be explicitly found or computed. To actually find and know the MPP, a technique known as the incremental conductance algorithm (ICA) can be used. In ICA, the ratio of voltage to current rather than their product is used to search for the MPP. By definition, the MPP occurs when:

$$\frac{dI}{dV} = \frac{-I}{V} \tag{6.1}$$

When dI/dV is greater than $-I/V$, the MPP is to the right of the current-operating point, at a larger current (Figure 6.4). When dI/dV is less than $-I/V$, the MPP is to the left of the current-operating point, at a smaller current.

The advantage of the ICA approach is that once the MPP is clearly identified, the operating point of the PV cell does not need to be perturbed or "shaken"

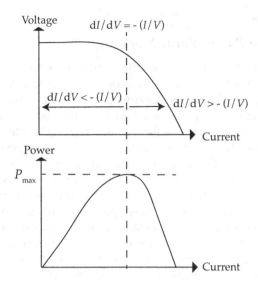

FIGURE 6.4
The incremental conductance algorithm (ICA) method for finding the maximum power point (MPP). In the ICA method, the relationship between the derivative of the current (with respect to the voltage) is compared to the ratio of current to voltage to determine in which direction the MPP lies from the present operating point.

anymore (Lokanadham and Bhaskar 2012). This reduces power losses incurred when the PV cell or cells are not operating at their maximum power and also reduces the power consumed in tracking the MPP. Furthermore, the ICA approach to MPPT performs better than the adaptive P&O approach when input lighting conditions change dramatically and frequently.

Many more accurate means of finding and maintaining the MPP in a PV cell or array of cells are available. Digital methods such as fuzzy logic and neural network control (Salah and Ouali 2011) and fractional voltage or current (Li, Zheng, Li, and He 2010) require knowledge of calibrated PV cell operating characteristics and reduce the amount of real-time information and computation required to compute the MPP. Such techniques also have a high degree of overhead and consume considerable power, which makes them better suited to traditional, grid-connected systems rather than smaller, wearable solar cell systems.

In portable and wearable solar cell systems, not only is the efficiency and power consumption of the MPPT system an issue, but size and weight of the overall system come into play. Analog implementations of MPPT are particularly attractive for portable and wearable solar cell systems because they can perform significant computing at lower power and smaller size than many corresponding digital systems.

MPPT that uses ripple correlation approaches is well matched to simple and inexpensive implementations using analog circuits. In ripple correlation, the time derivative of power in a PV cell or a small array of PV cells and the time derivative of current or voltage are used to push the power gradient to zero. Zero power gradient is by definition the point of maximum power production. Ripple correlation involves only small perturbations around the MPP, thus allowing the MPP to be found at speeds compatible with adapting to rapidly varying input conditions (Esram et al. 2006; Eltamaly 2018).

Regardless of how it is implemented, MPPT can be performed on a single PV cell, on a small subset of cells, or on an entire array of PV cells. Global MPPT ignores the MPPs of individual PV cells and instead tracks only the current and voltage of an entire array (Pallavee Bhatnagar and Nema 2013). While achieving more power output than in the absence of MPP tracking, this approach to MPPT does not guarantee that any one cell is operating at its MPP or that maximum power is extracted from a given string or other subset of PV cells in the array. Global MPPT, while space efficient and easy to implement, is only one step above no MPP tracking at all. On the other extreme, MPPT implemented at each individual cell level guarantees that maximum power is produced. However, the space and power consumed by individual cell MPPT is prohibitive for most systems. Regardless, in most solar cell systems, some type of MPPT is appropriate. Most MPPT is implemented at a substring level rather than a single cell or global MPP level. And, all techniques consider the operating conditions of the entire PV array and attempt to strike the best balance between power production and power and real-estate consumption incurred by the MPPT implementation itself.

6.4 Array Reconfiguration

Where irregular topologies are involved as is the case with many wearable systems, PV cells bend to do their jobs. Bending, twisting, and otherwise warping PV cells to anything other than their original flat footprint can quickly create significant power losses within series-connected strings caused by cells performing at very different operating points. This problem is partially mitigated by the fact that the penalty of shutting down a string submarined by underperformers is much less than in traditional, systems simply due to fewer cells per string. But, when power losses remain unacceptable under disparate operating conditions across a wearable array, reconfiguring the array is a possible solution.

Array reconfiguration (La Manna et al. 2014) allows PV cells in a system to operate at or near their individual MPPs. When PV cells in a string have differences in operating points that lead to unacceptable power losses, their cell connections can be electrically reconfigured so that they operate in strings with other more like-minded PV cells whose behavior is closer to their own. Static reconfiguration schemes allow connections to be reconfigured prior to installation to compensate for variations in the manufacturing process. Such one-time reconfiguration does not address or manage dynamic changes caused by partial shading, soiling, and other fluctuations in the environment that influence PV cell operation.

Dynamic physical array reconfiguration schemes offer even more opportunity for optimizing power production. These schemes allow for PV cells to be reconfigured and connections altered at any time during use to optimize power production and allow more individual PV cells to operate at or near their MPP. Connections are altered through the use of transistor-based switches that can adjust connections throughout the array on the fly. A wide variety of reconfiguration approaches have been demonstrated (La Manna et al. 2014). Common to all reconfiguration approaches is that they improve the power produced by a PV array that is subject to frequently and widely changing input conditions. Increases in power production occur at the expense of increased power consumption and greater system size incurred by the implementation of hardware switches, power-tracking electronics, and computation algorithms required for successful reconfiguration. A typical implementation of array reconfiguration uses a static array of PV cells and a dynamic bank of additional PV cells. PV cells from the adaptive bank are switched into or out of the static array as needed through a switching matrix to increase the overall power production. Making reconfigurable systems viable requires attention to reducing the complexity of the switching matrix and power-tracking electronics and also optimizing MPPT and reconfiguration algorithms. But, the best array reconfiguration systems continue to require substantial overhead, which may be justifiable for large arrays of PV cells that are typical of traditional installations but are prohibitive for wearable solar cell systems in all but the most power-hungry applications. For systems that do not have to reconfigure at high speeds, time domain adaptive reconfiguration (TDAR) is an option. TDAR reconfigures PV cells over time rather than space, allowing them to contribute to PV array power only when they are operating at comparable MPPs to other cells in the array (Vaidya and Wilson 2013). This approach accomplishes the same objectives as array reconfiguration based on switching matrices but does so without consuming the space and power required for such a matrix.

6.5 Summary

An immense amount of research and development effort has gone into improving the performance of solar panels in traditional systems. The design space around PV array design, array reconfiguration, mitigation of weak underperforming cells, and MPPT has been well explored. Wearable solar cell systems can and will benefit from these efforts.

As transistor technology advances, the scaling down of supply voltage can also go a long way to supporting wearable solar cell systems that perform better and cost less. If most of the power consumed by all of an individual's wearables and portables counted on a supply voltage of less than 1.5 V (rather than the 3.3 or 5.0 V that most wearables and smartphones require), PV systems could operate efficiently on strings containing only one or two cells. Designs using these smaller strings could greatly simplify array management strategies and the electronics required to support those strategies. While it is unrealistic to think that all wearables and portables could be scaled to these low-supply voltages, many can—particularly power-hungry microprocessors. Accordingly, while many of the lessons learned and technologies developed for traditional systems can be transferred to wearable solar cell systems, new design paradigms for these systems could lead to great leaps forward in the wearables and portables that can be seamlessly and shamelessly powered with ambient light.

References

Eltamaly, Ali M. 2018. "Performance of MPPT Techniques of Photovoltaic Systems under Normal and Partial Shading Conditions." In *Advances in Renewable Energies and Power Technologies*, 115–161. New York, NY: Elsevier.

Esram, Trishan, and Patrick L. Chapman. 2007. "Comparison of Photovoltaic Array Maximum Power Point Tracking Techniques." *IEEE Transactions on Energy Conversion* 22 (2): 439–449.

Esram, Trishan, Jonathan W. Kimball, Philip T. Krein, Patrick L. Chapman, and Pallab Midya. 2006. "Dynamic Maximum Power Point Tracking of Photovoltaic Arrays Using Ripple Correlation Control." *IEEE Transactions on Power Electronics* 21 (5): 1282–1291.

Falin, Jeff, and Wang Li. 2011. *A Boost-Topology Battery Charger Powered from a Solar Panel*. Texas Instruments Inc. http://www.ti.com/lit/an/slyt424/slyt424.pdf.

La Manna, Damiano, Vincenzo Li Vigni, Eleonora Riva Sanseverino, Vincenzo Di Dio, and Pietro Romano. 2014. "Reconfigurable Electrical Interconnection Strategies for Photovoltaic Arrays: A Review." *Renewable and Sustainable Energy Reviews* 33: 412–426.

Li, Weichen, Yuzhen Zheng, Wuhua Li, and Xiangning He. 2010. "A Smart and Simple PV Charger for Portable Applications." In *2010 Twenty-Fifth Annual IEEE Applied Power Electronics Conference and Exposition (APEC)*, 2080–2084. IEEE.

Lokanadham, Metta, and Kurba Vijaya Bhaskar. 2012. "Incremental Conductance Based Maximum Power Point Tracking (MPPT) for Photovoltaic System." *International Journal of Engineering Research and Applications (IJERA)* 2 (2): 1420–1424.

Pallavee Bhatnagar, A., and B. R. K. Nema. 2013. "Conventional and Global Maximum Power Point Tracking Techniques in Photovoltaic Applications: A Review." *Journal of Renewable and Sustainable Energy* 5 (3): 032701. https://doi.org/10.1063/1.4803524.

PV Education. 2019. "Module Circuit Design." https://www.pveducation.org/pvcdrom/modules-and-arrays/module-circuit-design.

Rawat, Rahul, and S. Chandel. 2013. "Hill Climbing Techniques for Tracking Maximum Power Point in Solar Photovoltaic Systems-a Review." *International Journal of Sustainable Development and Green Economics* 2: 90–95.

Salah, Chokri Ben, and Mohamed Ouali. 2011. "Comparison of Fuzzy Logic and Neural Network in Maximum Power Point Tracker for PV Systems." *Electric Power Systems Research* 81 (1): 43–50.

Samlex Solar. n.d. "The Difference Between Solar Cell, Module & Array." Accessed July 15, 2019. https://www.samlexsolar.com/learning-center/solar-cell-module-array.aspx.

Vaidya, Vaibhav, and Denise Wilson. 2013. "Maximum Power Tracking in Solar Cell Arrays Using Time-Based Reconfiguration." *Renewable Energy* 50: 74–81.

7

Energy Storage

In a majority of solar cell systems, the array of PV cells that transforms light into electrical energy does not directly produce a voltage or current that is compatible with the electronic devices and appliances that rely on it for power. Thus, one of the first tasks a solar cell system must perform for a PV array is to condition the output of the array to better match the requirements of subsequent electronic loads. In many solar cell systems, such conditioning involves two stages (Figure 7.1a). The first stage adjusts the DC voltage produced by the PV array to another DC voltage that is adequate to charge a battery or other energy storage device. This DC-DC conversion stage may also involve restricting the current that flows into the battery to prevent overcharging or storing power until sufficient current is available to optimally charge the battery. At this stage, the battery can supply electronic devices and appliances that run on DC. For other electronics, the second stage of many solar cell systems converts the DC output of the DC-DC converter to AC. This conversion allows the solar cell system to drive any AC appliances or devices that the battery cannot directly support. While some solar cell systems do not have a battery and supply power directly into a power grid, standalone systems and an increasing number of residential systems provide a battery for energy storage. In a wearable solar cell system, there is no need for a DC-AC inverter because almost all portables and wearables operate on DC (Figure 7.1b). With the exception of not needing a DC-AC inverters, the general structure of the wearable solar cell system is the same as many larger solar cell systems. After DC-DC conversion, additional charge controller electronics manage and adjust current and voltage to match the needs of the battery during different stages of charging. Because batteries have stringent charging requirements, some hybrid energy storage systems (HESS) use additional energy storage devices, such as supercapacitors, to enable batteries to charge closer to their optimal profiles. Advances in how PV array output is conditioned and how energy and power are stored continue to expand the devices, applications, and scenarios in which solar cell systems provide a viable solution to energy demand. This chapter focuses specifically on these parts of a solar cell system and how they can be designed to match the specific needs of wearable solar cell systems.

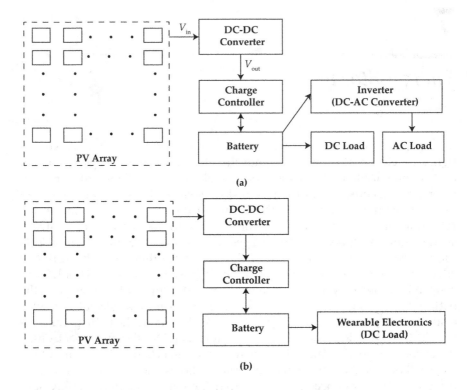

FIGURE 7.1
PV system designs. (a) PV systems designed to power both DC and AC devices and appliances and (b) PV systems for wearables and portables designed to power only DC electronics.

7.1 DC-DC Conversion

Even after an array of PV cells is designed and optimized for a particular application, the array output power, voltage, and current still vary with light irradiation in the ambient environment and with manufacturing mismatches, shading, aging, soiling, and other irregularities that cause the operating characteristics of individual PV cells to differ from one another. While the array management approaches discussed in Chapter 6 can allow PV cells to operate efficiently in response to a range of inputs, most batteries are not big fans of such variations and the fluctuations in current and voltage that result. DC-DC conversion circuits manage these fluctuations by stabilizing voltage.

Three fundamental approaches to DC-DC conversion are available for integration into solar cell systems (Table 7.1). The first assumes that the voltage provided by a PV array will always be lower than the voltage required

TABLE 7.1

Three Types of DC-DC Converters

Conditions	DC-DC Conversion
$V_{in} \leq V_{out}$	Boost
$V_{in} > V_{out}$	Buck
$V_{in} \leq V_{out}$ or $V_{in} > V_{out}$	Buck-boost

by the battery that is being charged. In this case, the function of DC-DC conversion will always be to increase or boost the PV array output voltage using a boost converter. The second approach assumes that the PV array voltage will always be higher than the desired charging voltage for the battery. A buck converter provides reduction in voltage during DC-DC conversion. And, finally when the input voltage range of the PV array overlaps with the range of desired battery charging voltages, a buck-boost converter is the right choice for DC-DC conversion.

The power losses inherent to each of these three types of DC-DC converters are an important consideration in the design of wearable solar cell systems. Power losses tend to be lowest and efficiencies highest among boost converters, with efficiencies approaching 99% in high-voltage/high-power implementations. Buck converters tend to exhibit lower efficiencies than boost converters overall although both experience more losses at lower input powers. And buck-boost converters have even lower efficiencies than buck converters and can experience greater losses at both high and low input powers (Graditi, Colonnese, and Femia 2010).

Wearable solar cell systems tend to power electronic devices with much lower voltage requirements than rooftop or other large-scale systems. The gap between PV array voltage and battery charging voltage is much smaller in wearable systems. For this reason, boost converters are often the best choice for these systems. However, given the highly variable power that wearable PV arrays generate as well as the wide range of loads that a wearable PV system might support, the best choice of DC-DC converter type and design should be optimized on a system-by-system basis.

7.1.1 Boost (Step Up) Conversion

A boost converter can be constructed with as few as four components (a switch, an inductor, a diode, and a capacitor), as shown in Figure 7.2a. The switch (S1) is frequently implemented using a metal oxide-semiconductor field effect transistor (MOSFET). When S1 is off, the switch can be modelled as an open circuit (Figure 7.2b) and when it is on, it can be modelled as a short circuit (Figure 7.2c).

For the boost converter to operate properly, S1 is turned on and off periodically with a duty cycle D, a frequency f (in Hz), and a period T (in sec)

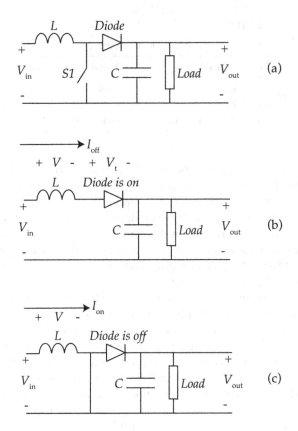

FIGURE 7.2
Basic boost converter design.

equal to $1/f$. The duty cycle (D) represents how long S1 is on relative to the period T:

$$D = \frac{On\ Time}{T} \tag{7.1}$$

When the boost converter is initially powered on, the output voltage V_{out} takes some time to reach its steady-state value which is a function of the input voltage (V_{in}). To understand what happens during this transient response, assume that V_{out} is initially zero. Assume also that the circuit begins with the input voltage abruptly increasing from zero to V_{in} and the switch S1 is off. In this situation (Figure 7.2b), current supplied by V_{in} flows through the inductor and the diode and charges the capacitor. If the diode is assumed to be ideal, the output voltage V_{out} charges to a value V_{in}. In practice, however, some voltage drop is experienced across the diode and after the initial period where S1 is *off*, V_{out} is less than V_{in} by an amount equal to the threshold voltage of the diode (V_t):

$$V_{out} = V_{in} - V_t \tag{7.2}$$

After a time $(1 - D) \times T$ seconds, the switch S1 turns on and the circuit operates, as shown in Figure 7.2c. The voltage across the diode decreases to a negative value equal to $0 - V_{out}$ which forces it to turn off, thereby conducting zero current. In this situation, the output voltage V_{out} has no available current path to ground and the capacitor at the output keeps the output voltage constant at $V_{in} - V_t$. In practice, the battery which is being charged by the boost converter presents a load resistance to the boost converter that discharges the output capacitor over time. To minimize this discharge, the capacitor is usually large which causes the decrease in output voltage to be negligible. At the same time, on the left-hand side of the boost converter, the voltage across the inductor changes abruptly from zero to V_{in}. The current through any inductor cannot change suddenly. Instead, the inductor current begins to gradually increase according to the governing relationship between current and voltage in an inductor:

$$V = L \frac{di}{dt} = L \frac{\Delta i}{\Delta t} \tag{7.3}$$

where V is the voltage across the inductor in volts (V), i is the current through it in amperes (A), and L is the inductance in Henries (H). While S1 is *on*, the inductor current seeks to increase from zero to the short circuit current of the PV cell or PV array that provides the input to the boost converter. By design, this short circuit current is never attained because it would damage or destroy the inductor, switch, or PV cells. In order to prevent such damage, the period T is short and the frequency f fast in order to restrict the maximum current that flows through any of the circuit components. Given a certain period T and an *on* time of $D \times T$, the maximum inductor current can be found as:

$$I_{max} = \frac{V_{in}}{L}(D \times T) \tag{7.4}$$

By the end of the *on* cycle of S1, the inductor current has reached the maximum value I_{max} and the switch returns to its *off* position (Figure 7.2b). Since inductor current cannot change suddenly, it remains temporarily at its maximum value as the switch S1 turns *off*. But, when the switch turns *off*, there is no place for the current to go—a frightening and impossible situation for the electrons in the circuit. To resolve the crisis, the voltage across the diode is forced to its threshold voltage V_t, thus causing it to turn on. Once the diode turns on, the output capacitor begins to charge once again, causing the voltage across it to increase from the value achieved during the first switching cycle $(V_{in} - V_t)$. During subsequent switching cycles (S1 *off* then S1 *on*), this process repeats itself, providing the mechanism by which V_{out} can continue to increase to a value that significantly exceeds the input voltage.

How much the output voltage is effectively "boosted" by the boost converter is determined by the duty cycle D and the period T. At steady state, the output voltage reaches a maximum value of:

$$\frac{V_{out}}{V_{in}} \sim \frac{1}{1-D} \tag{7.5}$$

This relationship can be derived by recognizing that, at steady state, the change in inductor current during the *off* period of the boost converter must be equal and opposite to the change in inductor current during the *on* period:

$$\Delta I_{on} = -\Delta I_{off} \tag{7.6}$$

The change in current during the *on* state (ΔI_{on}) is found by manipulating the governing equation for the inductor from derivative form (Equation 7.3) to integral form:

$$\Delta I_{on} = \frac{1}{L} \int_{t_o}^{t_1} V dt \tag{7.7}$$

The limits of integration begin at $t_o = 0$ sec and end at $t_1 = D \times T$ sec while the voltage across the inductor when the switch S1 is in the *on* position (Figure 7.2b) is simply V_{in}. Substituting into Equation 7.7 and solving the integral gives:

$$\Delta I_{on} = \frac{1}{L} \int_{0}^{DT} V_{in} dt = \frac{DT}{L} V_{in} \tag{7.8}$$

When the switch is *off*, Equation 7.7 can again be applied to determine the resulting change in inductor current. Since the switch turns off at time $t = D \times T$, the limits of integration now begin at $t_o = D \times T$ sec and end at $t_1 = T$ sec. The voltage across the inductor when the switch is in the *off* state is:

$$V = V_{in} - (V_{out} + V_t) \tag{7.9}$$

Integrating gives:

$$\Delta I_{off} = \frac{1}{L} \int_{DT}^{T} \left[V_{in} - (V_{out} + V_t) \right] dt = \frac{T}{L} \left[V_{in} - (V_{out} + V_t) \right] (1-D) \tag{7.10}$$

Setting the change in current ΔI_{off} during the *off* part of the cycle equal to the negative of the change in current during the *on* part of the cycle ΔI_{on} (Equation 7.6) in order to maintain steady-state conditions allows the output voltage to be determined:

$$\frac{T}{L}\left[V_{in} - (V_{out} + V_t)\right](1 - D) = -\frac{DT}{L}V_{in} \tag{7.11}$$

$$V_{out} = V_{in}\frac{1}{1 - D} - V_t \tag{7.12}$$

For high values of gain (V_{out} / V_{in}), the threshold voltage V_t is small compared to the first term on the right-hand side of the equation, and the gain given by Equation 7.5 is a good estimate of the actual gain. For wearable solar cell applications, however, where the gap between boost converter input and output voltages is typically smaller than in other solar cell systems, the threshold voltage of the diode cannot always be ignored and the expression for the output voltage given in Equation 7.12 is more appropriate.

Example: Consider a boost converter with input of 2 V, desired output voltage of 10 V, and maximum allowable inductor current of 1 A with a 1 mH inductor. Based on these constraints, the desired gain is 10/2 or 5, which requires a duty cycle of approximately 0.8 or 80%. If the maximum current of the inductor is 1 A, the required period of the clock is 0.625 msec, corresponding to a frequency of 1.6 kHz. The steady-state response of this boost converter is shown qualitatively in Figure 7.3. The boost converter output voltage is initially at 0 V and quickly ramps up, after several clock cycles to 10 V. Once at 10 V, the duty cycle maintains the voltage at 10 V. The inductor current increases while the switch S1 is *on* and decreases while switch S1 is *off*. In practice, the output voltage decreases when S1 is *on*, because there is a load resistance in parallel with the capacitor on the output voltage. Because the capacitor is typically large and the clock relatively fast, however, the discharge is relatively small.

The expressions for the gain provided in Equations 7.5 and 7.12 assume that the energy stored by the inductor while the switch S1 is *on* is not fully discharged (released) when the switch S1 is turned *off*. This assumption means that current will continue to flow during both *on* and *off* cycles of the switch (i.e., continuous operation). Under some conditions, particularly those involving a light load, the current in the inductor can reach zero before the switch transitions from *off* to *on*. This results in a discontinuous mode of operation that complicates the expression for the gain and causes it to be dependent on not only the duty cycle D but also the inductor value L, the clock period T, and the load resistance R_L that operates in parallel to the load capacitance C. Because of its complexity, higher output ripple, and lower efficiency, discontinuous operation is avoided in boost converter design for solar cell systems (Nelson 1986).

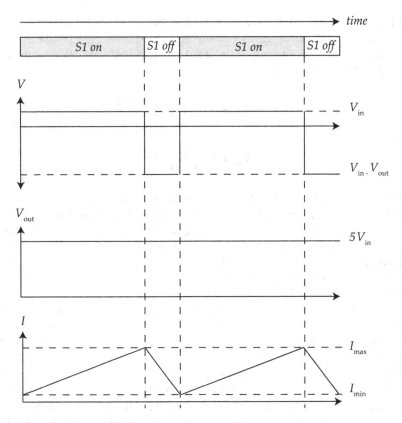

FIGURE 7.3
Boost converter behavior. The inductor current (I), inductor voltage (V), and output voltage (V_{out}) of the boost converter in continuous operation (where the inductor is not fully allowed to discharge between cycles) are shown. The output voltage is a multiple of the input voltage that is dependent on the duty cycle of the switch.

Whether the boost converter operates continuously or discontinuously, the overall goodness (i.e., figure of merit) of a boost converter is typically represented by its efficiency η:

$$\eta = \frac{P_{out}}{P_{in}} = \frac{P_{in} - P_{loss}}{P_{in}} \tag{7.13}$$

where the power lost in the boost converter itself arises from a variety of sources that are outside the scope of this discussion. While efficiencies close to 99% are possible in high-voltage/high-power boost converters, efficiency drops dramatically for light loads with maximum efficiencies around 90% for load currents less than 1 A (Davis 2018). Boost converter efficiencies also drop as the converter gain (V_{out} / V_{in}) increases. High gains are typically not an issue with wearable solar cell systems. For example, many solar-powered battery chargers contain only eight PV cells in series, capable of producing a maximum voltage of about 5.6 V (at 0.7 V per cell). A boost converter requires a gain of less than 3 to power an 11.1-V laptop computer (Falin and Li 2011).

The basic boost converter design, while simple, easy to implement, and low cost, has some disadvantages. A large capacitor is needed at the output to keep the output voltage stable and the ripple (i.e., periodic change) in the output current is significant. Furthermore, in conventional solar installations, where output voltages tend to be high, the basic boost converter design puts a large voltage stress on the MOSFET switch. While this last concern is minimal for the lower output voltage designs of wearable solar cell systems, a large ripple current can compromise optimal battery charging and a larger capacitor can increase cost and size of the overall solar cell system. An interleaved boost converter (Babba et al. 2018) can reduce the ripple on output power and current by using two boost converters operating out of phase with one another. This interleaved approach allows the output voltage to be charged for a longer period of time over each clock cycle, thus giving it less time to discharge. Lower discharge times also allow for a smaller capacitor to be used at the output of the boost converter. Other boost converter designs, that include isolation (of the output from the input), enhanced interleaved designs, and push-pull approaches (Li and He 2010; Babba et al. 2018), and address problems associated with high-power output demands and are less relevant to wearable applications.

Although generally less efficient than boost converters, buck converters are sometimes used in PV systems when the output voltage of the PV array is higher than the desired charging voltage of the subsequent battery under a majority of operating conditions (i.e., most of the time). The buck converter reduces the output voltage of a PV array to the required battery voltage. Like the boost converter, the duty cycle of a buck converter can be adjusted to maintain a stable output voltage when input conditions to the PV array fluctuate.

7.1.2 Buck (Step Down) Conversion

The basic buck converter design is shown in Figure 7.4. When the switch S1 is in the *on* state, the diode is off, no current flows through the diode as a result, and the input voltage supplies current to the inductor that subsequently

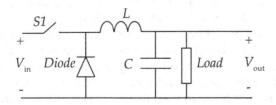

FIGURE 7.4
Basic buck converter design.

charges the load capacitor C. While the switch remains *on*, the inductor current continues to increase by an amount that can be calculated from the basic inductor relationship:

$$V_{in} - V_{out} = V_L = L\frac{\Delta I_L}{\Delta t} \tag{7.14}$$

where Δt is the time during which the switch S1 is *on* and is related to the duty cycle D and period T as $D \times T$. When the switch turns from *on* to *off*, the inductor current starts to decrease by an amount equal to the increase in current when the switch is *on* due to the conservation of power. In energy terms, the energy stored in the inductor while the switch is *on* is transferred to the output of the converter when the switch is *off*. A similar circuit analysis to the boost converter can be applied to the basic buck converter circuit to determine the gain of the buck converter:

$$\frac{V_{out}}{V_{in}} \sim D \tag{7.15}$$

Similar to the boost converter, the above expression assumes that the inductor current is continuous (i.e., not all of the inductor energy is discharged while the switch is off) and that the capacitor at the output is large enough that the output voltage (V_{out}) does not degrade significantly during the *off* part of the switching cycle. The basic buck converter has similar drawbacks to the basic boost converter and can be improved with more complex designs that are similar to those used to improve the boost converter. Despite their similarities, however, the buck converter is generally less efficient than the boost converter, with maximum efficiencies hovering around 97% for high-power applications and, like boost converters, are subject to degradation in efficiency at lighter loads and higher input voltages (Graditi, Colonnese, and Femia 2010).

7.1.3 Buck-Boost Conversion

In principle, a buck-boost converter can be used in a wearable solar cell system to accommodate any input voltage by boosting a low voltage to the requisite battery voltage and bucking or attenuating a high voltage. However, efficiency of the buck-boost converter degrades at both high and low voltages. This makes buck-boost converter performance incompatible with the highly variable output power and voltage generated by wearable solar systems. Thus, the best choice in wearable solar systems is often to use a boost converter and to design the PV array so that the input voltage will always be less than the desired output voltage.

7.2 Energy Storage

The primary design objective for a DC-DC converter in a wearable or portable solar cell system is to deliver a voltage and current that is matched to the needs of the energy storage device (e.g., battery, battery pack) that follows the DC-DC converter in the system architecture (Figure 7.1). While a wide range of battery technologies are available to support energy storage in PV systems, a majority of consumer electronics and other wearable devices rely on only one of three materials and technologies: nickel cadmium (NiCd), nickel metal hydride (NiMH), or lithium-ion (Li-ion). In contrast, when requiring energy storage, stationary solar cell systems have historically relied on lead-acid batteries.

7.2.1 Batteries

One of the more common indicators of the goodness of a battery, whether rechargeable or single-use, is its capacity, which is typically provided in ampere-hours (Ah) or milliampere-hours (mAh). Unfortunately, there are a number of drawbacks to using capacity to identify battery quality. First, capacity in Ah or mAh speaks to the amount of current that can be delivered over a certain period of time. Ideally, a 20-mAh battery would deliver 20 mA for 1 hour or 10 mA for 2 hours. But neither scenario speaks to the true energy capacity of the battery. For this, the capacity must be multiplied by the nominal voltage of the battery. The nominal voltage, in turn, is dependent on the battery material and technology. For example, a NiCd battery has a nominal cell voltage of 1.2 V. With a capacity of 10 mAh, the battery can deliver 0.012 Wh. In contrast, a Li-ion battery with a cell voltage of 3.6 V and the same capacity, can deliver three times that amount or 0.036 Wh. Furthermore, capacity typically degrades with increasing current demand and does so differently depending on the type and size of the battery as well as environmental factors including temperature. For portables and wearables, these issues are typically addressed during the design of electronic devices and not during the design of charging systems, whether these charging systems are based on AC, DC, solar, or some other energy source.

Of more concern for wearables and portables is the size and weight of the battery, as the battery is often the most cumbersome component in these devices. Specific energy and energy density provide insight into how cumbersome (or not) a battery can be in a wearable or portable application. Specific energy, quantified in watt-hours per kilogram (Wh/kg) provides an estimate of how much a battery will weigh (not including electronics required to control charging and ensure safety). To gauge the relative size, the energy density provides a measure of nominal energy generated per unit volume (Wh/L). These parameters are largely a function of the battery

TABLE 7.2

Size and Volume of Batteries for Solar Cell Systems

Material	Specific Energy	Energy Density	Size/Weight Suitability for Wearables and Portables
Lead-Acid	35–40 Wh/kg	80–90 Wh/L	Poor
Lithium-Ion (Li-ion)	90 Wh/kg	210 Wh/L	Excellent
Nickel Cadmium (NiCd)	50 Wh/kg	140 Wh/L	Good
Nickel Metal Hydride (NiMH)	55 Wh/kg	180 Wh/L	Good

Source: Simpson (2011) and May, Davidson, and Monahov (2018).

material and technology with Li-ion leading the way in modern portable and wearable devices (Table 7.2).

There is more to the goodness of a battery than how big it is and how much it weighs. Until the mid to late 1990s, NiCd was the only appropriate battery technology for portable electronic devices. These batteries were regularly used in emergency radios, video cameras, and power tools. As early as 1950, advances in NiCd battery technology led to the familiar sealed NiCd battery pack, and further advances in technology led to ultrahigh capacity versions of the NiCd battery. NiCd batteries have relatively low cell voltages (1.2 V), however, and multiple batteries must be stacked in series in order to power most portable and wearable devices. These low cell voltages are also more compatible with the output of PV cells. But, memory effects, where the battery must fully discharge in order to retain its full capacity, as well as high self-discharge when the batteries are not in use, limit their use in seamless, wearable solar cell systems. A unique advantage of NiCd batteries and battery packs, which is valuable in certain electronic devices, is that the cell voltage remains at approximately 1.2 V until right before the battery becomes dead or unusable. This behavior is in sharp contrast to most other batteries whose cell voltage degrades more gradually during discharge.

With the same 1.2 V per cell, NiMH batteries have higher capacity than NiCd and fewer memory effects. These batteries are also environmentally friendly and operate over a wide temperature range with greater capacity than NiCd. However, NiMH batteries are difficult to charge, vulnerable to overheating, sensitive to overcharge, and like NiCd, discharge quickly when not in use. These disadvantages typically eliminate NiMH batteries from consideration in most wearables and portables.

Lithium has long been considered an attractive alternative to NiCd and NiMH batteries. But, despite the fact that work on lithium batteries began in the early 1900s, safety issues prevented the technology from maturing for almost a century. Finally, in the 1990s, lithium battery technology in the form of lithium-ion (Li-ion) chemistries became sufficiently viable and safe to compete with NiCd. Since the 1990s, lithium-based battery technologies have advanced sufficiently to become the fastest-growing and most promising battery technology for the future of powering portable devices. Lithium is very attractive for wearable and other mobile applications because as the lightest of metals, lithium offers the highest energy density, most lightweight battery, and highest electrochemical potential among its competitors. Because lithium is unstable, a nonmetal alternative (Li-ion) with well-managed control of both charging and discharging cycles has enabled the advantages of lithium to be exploited safely. Li-ion battery cells also have high single-cell voltages (approximately 3.6 V), which make single-cell batteries viable for wearable solar cell systems.

In consideration of specific energy, energy density (Table 7.2), and other features and drawbacks of each of these battery technologies (Table 7.3), lead-acid emerges as the most reliable choice for stationary standalone PV systems, whereas Li-ion is often the best choice for mobile and wearable solar cell systems. Managing the charging of these batteries with PV arrays is relatively straightforward for lead-acid batteries but gets increasingly complicated for Li-ion and other newer battery technologies.

While PV arrays are inherently ill-suited to delivering a stable current and voltage because of changing illumination and array conditions, control circuits can be designed into a wearable solar system to minimize the impact of such high variability on battery performance. These control circuits and

TABLE 7.3 Batteries for Standalone PV Systems

Material	Lead-Acid	Li-ion	NiCd	NiMH
Nominal voltage per cell	2 V/cell	3.6 V/cell	1.2 V/cell	1.2 V/cell
Recharging lifetime[a]	200–300 cycles	500–1,000 cycles	1,500 cycles	300–500 cycles
Self-discharge rate[b]	5%/month	5–10%/month	15–20%/month	20–30%/month
Charging time[c]	5–8 hours	1–2 hours	4–10 hours	12–36 hours
Cost	$100–$200/kWh	$300-$1,000/kWh	$300–$600/kWh	$300–$600/kWh
Environmental toxicity	High	Low	High	Low
Advantages	Very inexpensive	Lightweight, small	Stable cell voltage	Environmentally friendly
Disadvantages	Heavy, large	Can be unstable, and natural resource extraction is difficult	Demonstrates high self-discharge and has memory effects	Demonstrates high self-discharge and is difficult to charge

Source: Simpson (2011) and Battery University (2017a, b, 2018).
[a] Number of recharging cycles at which battery degrades to 80% of original capacity.
[b] Includes energy drain incurred by safety circuits.
[c] Given as slow charging time; fast charging time requires more complex circuitry and poses safety hazards.

corresponding strategies can vary with battery type, because different battery technologies have different needs for optimal charging.

As a point of departure for understanding how batteries are best charged, consider the lead-acid battery around which many stationary solar cell systems have been designed. Lead-acid batteries are a relatively inexpensive, robust, reliable, and mature technology. Their specific energy is lower than NiCd, NiMH, and Li-ion batteries (Table 7.2), but their low cost keeps the technology in play for utility-scale and other large-scale energy storage systems. Lead (Pb) acid battery technology itself is considered more sustainable than competing technologies in terms of resource availability, and a vast majority of lead batteries are recycled, which further improves sustainability and reduces the end-of-life impacts on environmental and ecosystem health (May, Davidson, and Monahov 2018). However, Pb batteries contain a toxic heavy metal and are significantly heavier than NiMH, NiCd, and Li-ion technologies, they are a nonstarter in most wearable PV applications. Nevertheless, the maturity and success of lead-acid in many solar cell systems provides fundamental insight into how to overcome challenges involved in charging a battery using the inconsistent and often unstable output that PV arrays often provide.

A typical, optimized charging cycle for a lead-acid battery consists of three phases (Figure 7.5): (a) *bulk* charge, (b) *tapered* charge, and (c) *float* or *trickle* charge. Initially and for most of its charging cycle, a constant current

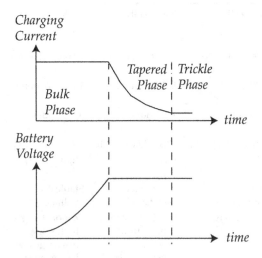

FIGURE 7.5

Three basic phases of battery charging. A standard rechargeable battery like lead-acid is first charged using a constant current (the bulk phase). When a certain voltage is achieved, the charging controller maintains a constant voltage and the current through the battery drops as the battery continues to charge (tapered phase). When the battery is fully charged, the charging controller continues to inject a small amount of current into the battery to offset self-discharge of the battery (trickle phase).

is provided to *bulk* charge the battery. Once the battery voltage reaches a predetermined value, the charging current enters a *tapered* charge phase where it gradually declines while holding the battery voltage constant. After some time when the battery is fully charged, the charging voltage is reduced or removed to avoid damaging the battery via excessive heating or loss of electrolyte. At this time, the battery enters into a *trickle* or float phase where only a small amount of trickle current flows into the battery, enough to offset its self-discharge behavior. *Float* charging includes trickle charging but also adds additional protection to avoid overcharging and subsequent damage to the battery (Ross 2003).

The challenges involved in adapting a solar cell system to these three basic phases are nontrivial. As discussed in the previous section, the duty cycle of a boost or buck converter can be adjusted to control the voltage input to the battery, but that voltage must be converted to a current to adapt to these three phases of battery charging. Two basic designs can be used to adjust the charging current (Figure 7.6). In both designs, a diode is used to prevent current from flowing backward from the battery to the DC-DC converter and the PV array. The battery charger may also contain a series regulator (Figure 7.6a) that behaves as a variable resistance in series with the output of the PV array or DC-DC converter, decreasing or increasing the current into the battery by increasing or decreasing the series resistance. Or, alternatively, a shunt regulator (Figure 7.6b) can be used to divert excess current that flows out of the PV array or DC-DC converter, with the amount of diverted current determined by the controller. Unfortunately, implementing these shunt or series regulators as variable resistors using either resistors or transistors

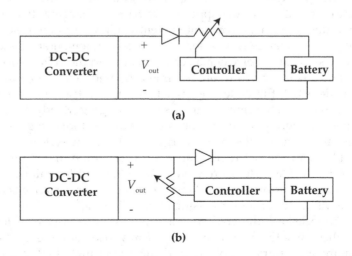

FIGURE 7.6
Basic charge controller architectures. Current can be reduced by (a) a series resistor or (b) a shunt resistor to maintain more constant current during charging. Unfortunately, both architectures result in power loss when regulating current in this way.

often consumes a prohibitive amount of power and is impractical in many wearable solar cell systems. Instead, a transistor, implemented as a switch, can be operated at different duty cycles to establish an average current into the battery that is as close as possible to the optimal charging current for the battery, whether in bulk, tapering, or trickle/float phases of charging. In addition to adjusting the duty cycles of the shunt or series regulation, the controller is also designed to discontinue charging the battery or to maintain only a trickle charge once the battery has fully charged (Ross 2003).

Unfortunately, lead-acid batteries are almost entirely impractical for wearable solar cell systems. And, the practical alternatives to lead-acid (Li-ion, NiCd, NiMH) that power most wearable, portable, or mobile electronic devices have more complex charging needs. For instance, Li-ion batteries, while using a constant current/constant voltage (cc-cv) charging strategy that is similar to that used in a lead-acid battery, have some special needs. The voltage of Li-ion batteries is often capped or limited to a voltage that is less than the maximum or saturation voltage of the battery. Limiting the battery voltage in this way prevents complete recharging, often capping the charge at about 85% of its maximum capacity. The benefit of this approach is that it prevents the battery from getting overstressed and preserves its lifetime (i.e., recharge cycles). Furthermore, the unstable nature of Li-ion batteries requires specialized protection circuitry to discontinue or adjust charging when the temperature rises too far above the ambient temperature. To their advantage, Li-ion chargers do not require a trickle or float phase, because their self-discharge rate is very low and limited to the current demanded by such protection circuitry (Reddy 2011).

Nickel-based batteries (e.g., NiCd, NiMH) do not use this constant current, constant voltage (cc-cv) approach to battery charging. Instead, in an ideal charging scenario, a constant current is allowed to increase the voltage freely. A full charge is detected when a certain temperature increase is detected (less accurate) or when a certain change in the voltage rise or voltage signature is identified (more accurate). As with lead-acid, when a full charge is detected, the battery is switched into the trickle or float charge mode to avoid overcharging and to offset discharging (Reddy 2011).

In general, batteries are an inexpensive means to store energy (compared to nonchemical means of storing energy) and have low self-discharge rates that enable them to hold their charge for a long time under zero or light loads. Rechargeable batteries rely on a moderate load to maintain capacity and discharging performance and require a stable energy source at the input to charge well without compromising battery lifetime.

While batteries are the gold standard in storing energy over long time periods, they have their disadvantages. Heavy loads at the output of a battery require short bursts of power and correspondingly high currents that can cause dramatic drops in capacity and discharge time. In many solar cell systems, the power density limitations of the battery, regardless of its energy density, can limit its usefulness as an energy storage device. For example, the lead-acid battery can deliver 35–40 Wh/kg of battery weight, but has a power density of only 10–20 W/kg (Committee on Soldier Power/Energy Systems, Board on Army Science and Technology, and National Research Council 2004). This means that a 1-kg battery can deliver a maximum instantaneous power of only 10–20 W. But, many large-scale solar cell systems are often used to power motors, electric vehicles, and other devices that require high intermittent currents. Worst case, a battery cannot provide these currents at all due to limitations in power density. Best case, battery capacity is compromised by such high current demand, leading to a shorter and undesirable discharge time and shortened useful life.

On the other side of the battery, unstable fluctuations in the energy source used to recharge the battery can compromise charging effectiveness. At best, such fluctuations result in charging behavior that is not well matched to the needs of the battery and also decrease the lifetime of the battery as measured in the number of recharge cycles. At worst, a poor recharge profile can cause permanent damage to the battery itself. Many large stationary solar cell systems suffer the most from fluctuations in the battery load caused by the number, type, and behavior of devices connected to the system. In contrast, wearable solar cell systems suffer less from large variations in load and more from fluctuations in input light energy and electrical power provided to the battery. Not only must wearable solar cell systems be able to harvest energy from many types of light sources at a wide range of irradiation (intensity levels), but they must also tolerate frequent, and rapid changes in input light conditions.

7.2.2 Supercapacitors

Supercapacitor technology has the potential to mitigate incompatibilities between the stable charging profiles that batteries require and the unstable voltage and current that PV arrays often provide. While a battery relies on chemistry to store energy, supercapacitors rely on an electric field to store energy. Energy is stored via the separation of negative and positive charges onto two, separate conductive plates. The two plates are separated by a dielectric or insulating material, which prevents the charges from crossing the gap between the plates.

Traditional capacitors are often limited to values of less than a single Farad, but supercapacitors can be manufactured to deliver thousands of Farads per gram, thousands of charging cycles, and power densities of over 100,000 Wh/kg (Sharma, Arora, and Tripathi 2019). Supercapacitors can also be charged very quickly compared to batteries without many of the negative consequences of rapid charging that batteries experience (Battery University 2019). Supercapacitors are not a practical option to replace batteries, however, because they cost more than conventional batteries. Costs of supercapacitors are approximately $1,000/kWh while lead-acid and Li-ion batteries cost as little as $100 and $300/kWh, respectively (Battery University 2017a). For economic reasons alone, then, supercapacitors are unlikely to ever fully replace batteries. Supercapacitors also have substantially higher self-discharge rates than batteries, with initial discharges measurable in hours and secondary discharges occurring on the order of days (Kowal et al. 2011). Despite these disadvantages, supercapacitor technology can and has been used to complement battery technologies in hybrid energy storage systems (HESS). In these HESS, bidirectional DC-DC converters are used to allow a PV array to charge a battery, a supercapacitor, or both in order to drive DC loads associated with many electronic devices. The supercapacitor is often placed in parallel with the battery to supply a boost in current and power to the output load on an as-needed basis. HESS approaches have combined supercapacitors with valve-regulated lead-acid and other types of batteries for many stationary solar cell systems (Glavin and Hurley 2012; Kollimalla, Mishra, and Narasamma 2014; Ma, Yang, and Lu 2015), microgrids (Zhou et al. 2011), wireless sensor networks (Ongaro, Saggini, and Mattavelli 2012), and other applications where producing electricity from light energy is viable.

While wearable solar cell systems in general do not power heavy DC loads or loads prone to temporary surges of high current or power, they do suffer from unusually high variability in input irradiation. Light impinging on a wearable solar cell system can vary from high-energy direct sunlight to indirect artificial lighting and, on a mobile platform, can vary as frequently as second to second. General purpose, wearable solar cell systems designed to power a wide range of consumer devices can benefit significantly from HESS that supplement battery storage with an alternative energy storage device such as the supercapacitor. Without an alternative form of energy storage, these systems will be forced to discard power that does not meet battery requirements or may cause batteries to prematurely reach the end of their useful life.

7.3 Summary

The output of a PV array is generally insufficient to provide power directly to electronic devices and appliances. Like many forms of renewable energy, solar power generated from PV cells and arrays is neither consistent nor always predictable. Even when operated at its maximum power point (MPP), a PV array produces currents that drop dramatically with decreasing irradiation (i.e., input light). While PV array voltage does not change nearly as much as PV array current, it still experiences fluctuations as the MPP shifts with changing input conditions. In order to accommodate this variability and provide stable power to electronics in wearable solar cell systems, the voltage output of the PV array must first be adjusted to meet the power supply needs of the electronics. This is done through the use of a boost, buck, or buck-boost converter that converts the DC input voltage to a more desirable DC output voltage. The input-output characteristic of the DC-DC converter is often adjusted to maintain a nearly constant, stable output voltage. Once the voltage is stabilized, the current from the PV array and the DC-DC converter is adjusted to best meet the needs of the energy storage device in the system. Adjustments in current can be made through shunt or series regulators, but power losses using these approaches can be significant. Alternatively, super-capacitors and other energy storage devices can be used to supplement batteries in order to more closely approach ideal charging profiles and reduce the amount of power that is ultimately dissipated or discarded to maintain these profiles. Thus, while individual PV cell efficiency is often at the forefront of solar system design, the choice of interface electronics and energy storage strategy can also make a substantial difference in improving the performance of the overall system.

References

Babba, S. E., G. El Murr, F. Mohamed, and S. Pamuri. 2018. "Overview of Boost Converters for Photovoltaic Systems." *Journal of Power and Energy Engineering* 6: 16–31.

Battery University. 2017a. "BU-1006: Cost of Mobile and Renewable Power." https://batteryuniversity.com/learn/article/bu_1006_cost_of_mobile_power.

Battery University. 2017b. "What's the Best Battery?" https://batteryuniversity.com/learn/archive/whats_the_best_battery.

Battery University. 2018. "BU-802b: What Does Elevated Self Discharge Do?" https://batteryuniversity.com/learn/article/elevating_self_discharge.

Battery University. 2019. "BU-209: How Does a Supercapacitor Work." April 17, 2019. https://batteryuniversity.com/learn/article/whats_the_role_of_the_supercapacitor.

Committee on Soldier Power/Energy Systems, Board on Army Science and Technology, and National Research Council. 2004. *Meeting the Energy Needs of Future Warriors*. Vol. Appendix C. Washington D.C.: National Academies Press. https://doi.org/10.17226/11065.

Davis, Nick. 2018. "High Efficiencies at Light Loads: A Voltage Boost Converter from Texas Instruments – News." April 12, 2018. https://www.allaboutcircuits.com/news/high-efficiencies-at-light-loads-voltage-boost-converter-Texas-Instruments/.

Falin, Jeff, and Wang Li. 2011. *A Boost-Topology Battery Charger Powered from a Solar Panel*. Texas Instruments Inc. http://www.ti.com/lit/an/slyt424/slyt424.pdf.

Glavin, M. E., and W. G. Hurley. 2012. "Optimisation of a Photovoltaic Battery Ultracapacitor Hybrid Energy Storage System." *Solar Energy* 86 (10): 3009–3020. https://doi.org/10.1016/j.solener.2012.07.005.

Graditi, G., D. Colonnese, and N. Femia. 2010. "Efficiency and Reliability Comparison of DCDC Converters for Single Phase Grid Connected Photovoltaic Inverters." In *Proc IEEE SPEEDAM*, 140–147.

Kollimalla, S. K., M. K. Mishra, and N. L. Narasamma. 2014. "Design and Analysis of Novel Control Strategy for Battery and Supercapacitor Storage System." *IEEE Transactions on Sustainable Energy* 5 (4): 1137–1144. https://doi.org/10.1109/TSTE.2014.2336896.

Kowal, Julia, Esin Avaroglu, Fahmi Chamekh, Armands Šenfelds, Tjark Thien, Dhanny Wijaya, and Dirk Uwe Sauer. 2011. "Detailed Analysis of the Self-Discharge of Supercapacitors." *Journal of Power Sources* 196 (1): 573–579.

Li, W., and X. He. 2010. "Review of Nonisolated High-Step-Up DC/DC Converters." *IEEE Transactions on Industrial Electronics* 4: 1239–1250.

Ma, Tao, Hongxing Yang, and Lin Lu. 2015. "Development of Hybrid Battery–Supercapacitor Energy Storage for Remote Area Renewable Energy Systems." *Applied Energy* 153: 56–62. https://doi.org/10.1016/j.apenergy.2014.12.008.

May, Geoffrey J., Alistair Davidson, and Boris Monahov. 2018. "Lead Batteries for Utility Energy Storage: A Review." *Journal of Energy Storage* 15 (February): 145–157. https://doi.org/10.1016/j.est.2017.11.008.

Nelson, Carl. 1986. "LT1070 Design Manual: Application Note 19." Linear Techology. https://www.analog.com/media/en/technical-documentation/application-notes/an19fc.pdf.

Ongaro, F., S. Saggini, and P. Mattavelli. 2012. "Li-Ion Battery-Supercapacitor Hybrid Storage System for a Long Lifetime, Photovoltaic-Based Wireless Sensor Network." *IEEE Transactions on Power Electronics* 27 (9): 3944–3952. https://doi.org/10.1109/TPEL.2012.2189022.

Reddy, Thomas B. 2011. *Linden's Handbook of Batteries*. Vol. 4. New York: McGraw-Hill.

Ross, J. Neil. 2003. "System Electronics." In *Practical Handbook of Photovoltaics: Fundamentals and Applications*. Boca Raton, Florida: CRC Press.

Sharma, Kriti, Anmol Arora, and S. K. Tripathi. 2019. "Review of Supercapacitors: Materials and Devices." *Journal of Energy Storage* 21: 801–825.

Simpson, Chester. 2011. "Characteristics of Rechargeable Batteries." Texas Instruments Literature Number SNVA533. http://www.ti.com/lit/an/snva533/snva533.pdf.

Zhou, H., T. Bhattacharya, D. Tran, T. S. T. Siew, and A. M. Khambadkone. 2011. "Composite Energy Storage System Involving Battery and Ultracapacitor With Dynamic Energy Management in Microgrid Applications." *IEEE Transactions on Power Electronics* 26 (3): 923–930. https://doi.org/10.1109/TPEL.2010.2095040.

8

Wearable and Portable Technology

The electronic devices that individuals in tech-savvy societies carry on their person are many, varied, and hungry for power. More and more, general-purpose portable devices such as laptop computers, tablets, and smartphones are not able to survive the time that they spend away from a wall outlet or other stationary recharging station. As portable power banks expand to accommodate expectations for convenient recharge of portable computers, phones, and similar devices, they also increase the environmental burden that batteries place on an already limited global supply of natural resources. While often far less power hungry than general-purpose portables, wearables nevertheless add to this mobile energy demand.

Traditional solar cell systems seek to supply substantial amounts of power while displacing non-renewable sources of electricity. In so doing, they play a critical role in curbing air and water pollution, greenhouse gas emissions, and other environmental costs associated with coal, natural gas, and other nonrenewables. In contrast, wearable solar cell systems are not tasked with generating large amounts of power. Rather, their primary design focus is on delivering more reliable, convenient, and consistent power to the user while also mitigating the environmental impacts imposed by the proliferation of batteries. In the future, they can also open the door to wearables and portables that are not presently viable because of limitations in the availability of portable power. Whether or not and to what degree wearable solar systems can accomplish these goals begins with understanding the present scope of wearables and portables with particular attention to the energy and power that these devices demand.

8.1 Mobile Phones

Mobile phones, particularly smartphones, will continue to be ubiquitous for years to come. By 2021, Cisco Systems predicts that more people in the world will use a mobile phone than will have bank accounts or access to running water (Franklin 2017). In 2019, 94% of adults in advanced economies owned a mobile phone, while 83% in emerging economies owned one.

A majority of these phones are smartphones with 76% and 45% of people owning these mobile marvels in advanced and emerging economies, respectively (Pew Research Center 2019a). The smartphone is not only more widely used than the traditional mobile phone but is also the most commonly owned mobile electronic device on the planet (Pew Research Center 2019b). Because of its enhanced features and user connectivity, the smartphone consumes significantly more power than traditional flip-style phones.

Regardless of the type of phone they own, most consumers charge their phones at least once a day and some in major cities like New York and Philadelphia do so over twice a day (Veloxity 2017). Consistent with this level of usage, the daily energy demand for a typical smartphone user is estimated in Table 8.1 for multiple models of smartphone from four different manufacturers and a single model from each of four manufacturers of traditional flip phones. Estimates of energy demand are provided in watt-hours per day (Wh/day) and assume that the phone is fully recharged once per day and that each battery's watt-hour (Wh) rating can be estimated as the product of its published ampere-hour (Ah) (1,000 × mAh) capacity and the voltage of a standard lithium-based battery (3.7 V). Daily energy demand estimates vary widely even among phones made by the same manufacturer. For example, different models of the Apple iPhone consume between 6.35 and 11.74 Wh/day.

TABLE 8.1

Energy Demand of Mobile Phones

Smartphone	Battery Capacity (mAh)			Daily Energy Demand (Wh)[a]		
	Min.	Mean	Max.	Min.	Mean	Max.
Smartphones						
Apple iPhone	1,715	2,092	3,174	6.35	7.74	11.74
Motorola Moto	3,760	4,380	5,000	13.91	16.21	18.50
OnePlus	3,000	3,333	3,700	11.10	12.33	13.69
Samsung Galaxy	1,900	2,892	3,600	7.03	10.70	13.32
Average		3,174			11.75	
Traditional Mobile (Flip) Phones						
Convoy		1,300			4.81	
Gusto		1,000			3.70	
LG		950			3.52	
Razer		780			2.89	
Average		1,008			3.73	

Source: Rioja (2019) and Grush (2019).

[a] Assumes user fully recharges the phone once per day and the battery is rechargeable lithium with a cell voltage of 3.7 V.

8.2 Other Portables

Smartphones are not the only devices that many individuals carry for general-purpose connectivity and computing. For instance, over half of the adults in the United States own a tablet (Statista 2019), over 70% own a laptop computer, almost 20% own an e-book reader, and 40% own an MP3 player (Pew Research Center 2015). Worldwide, personal computer ownership is approaching 50% with laptop computers dominating over desktop computers (Statista 2018). While both the ownership and use of these other portables may not be as ubiquitous as that of mobile phones, most portable computing devices make up for less usage by consuming significantly more power than mobile phones. Energy demand for several popular portables is estimated in Table 8.2 by taking the ratio of 90% of battery capacity (in watt-hours) to battery life as reported from third-party testing. Most portables do not fully discharge but instead shut down or go to sleep when approximately 10% of the battery remains. Thus, battery life covers only 90% rather than 100% of total battery capacity. As estimated in Table 8.2, average energy usage over the life of a battery in watt-hours per hour is the same as average power consumption (in watts).

TABLE 8.2

Energy Demand of Portable Computing Devices

Device	Battery Capacity (Wh)	Battery Life (h)	Energy Demand[a] (Wh/h)
Laptop Computers			
Lenovo Thinkpad T480	24	17	1.3
HP Envy X2	25	14	1.6
Dell XPS 13 9370	52	12	4.0
Apple MacBook Pro (15 inch)	84	12	6.3
Tablet Computers			
Apple iPad Pro (10.5 inch)	30	14	2.1
Lenovo Yoga Book	34	9	3.8
Samsung Galaxy Tab S3	23	9	2.6
Asus ZenPad 3S 10	18	8	2.3
Microsoft Surface Pro	45	6	7.5
MP3 Players			
Apple iPad Touch	3.4	40	0.085
SanDisk Clip Jam	1.1	18	0.061
Sony Walkman NW-WS413	1.4	12	0.12

Source: Osborne (2017), Reisinger (2019), and Tabari (2019).

[a] Energy demand is estimated as the ratio of 90% of battery capacity to battery life, is given in Wh/h, and is roughly the same as average power consumption in watts.

E-readers are also popular portables, although sales continue to decline (Statista n.d.). These devices have similar battery capacity and energy demand during continuous use as other types of portable computing devices. For example, a recent model of the top-selling Kindle e-reader, the Kindle Paperwhite made by Amazon has a 1,420-mAh battery capacity (5.3 Wh assuming a 3.7 battery voltage) and has an anticipated battery life of 6 weeks at 30 min/day (equivalent to 21 hours of continuous use) with Wi-Fi disabled. This amounts to an approximate energy demand of 0.25 Wh/h (i.e., an average power consumption of 0.25 W) when the device is in use. Not only does this put the e-reader at a much lower level of average power consumption than the laptop computer, the lower device usage associated with e-readers amounts to a daily energy demand (0.125 Wh/day assuming 30 minutes of usage per day) that is about 1% of what a smartphone demands.

8.3 Wearable Devices

Wearable devices come in many shapes, sizes, and functions and are sold all over the world. The market for wearable technology is projected to grow to over $70 billion by 2022 (Grand View Research 2016) with North America accounting for almost 50% of wearable technology revenue. Fitness and wellness account for the largest portion of wearable technology sales followed by healthcare, infotainment, defense, and enterprise/industrial applications (Grand View Research 2016). Most wearable technology is electronic, consumes a nontrivial amount of power, and requires an energy storage device like a battery to operate properly. The variations in the amount and type of power and energy required by wearables creates many opportunities for wearable solar cell systems to meet the energy demands of these devices.

Of the many wearables on the market, four are expected to dominate the worldwide market by 2022. Over 158 million ear-worn devices are expected to be sold, 193 million wrist-worn devices including smartwatches, sports watches, and smart wristbands, 80 million head-mounted displays, and almost 20 million pieces of smart clothing (Gartner 2018). However, the options for wearables extend well beyond the head and the wrist (Figure 8.1). While a comprehensive overview of the many shapes and sizes of wearables is outside the scope of this chapter, devices chosen to represent the full range of possibilities are explored to gain an appreciation for the need for wearable energy harvesting and storage.

8.3.1 On the Ears

Ear-worn devices span many product sectors for wearables and include popular products such as Apple's AirPods, Samsung's IconX, and Plantronics'

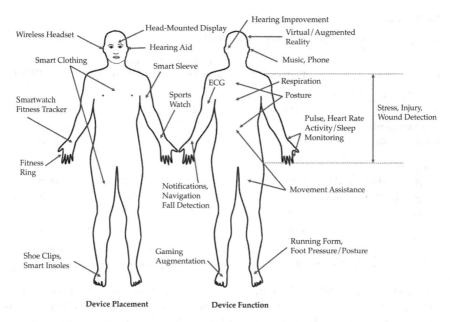

FIGURE 8.1
Wearables anyone?

BackBeat FIT (all wireless headsets) that are expected to hold about 30% of the world market share in the near future (Gartner 2018). AirPods are a communication wearable which allow bidirectional communication between the user and a wide variety of Apple devices including the iPhone, iPad, and AppleWatch. AirPods contain both optical and motion sensors that limit audio play to time when the user is wearing the AirPods and also take tap commands to control music and other auditory playback. The ear-worn AirPods also contains an accelerometer for speech detection and noise reduction (Apple n.d.). Samsung's IconX headset provides both music player capability (with or without any accompanying smartphone nearby) and fitness tracking capability to log time, distance, and calories burned during walking or running (Samsung n.d.). And, Plantronics BackBeat FIT acts as a music player with a stopwatch and related functionality including an eartip that allows ambient sound to enter the ear canal in order to support the user's situational awareness and safety (Plantronics n.d.).

Beyond these general-purpose products among ear-worn wearables, hearing aids account for about 15 million (The Hearing Review 2018) or 45% of the estimated 33 million ear-worn units sold worldwide in 2018 (Gartner 2018). About 75% of hearing aids are of the behind the ear style (Strom 2018) and most use single-use, button-type batteries. In contrast, a majority of wireless headsets or headphones including AirPods, IconX, and BackBeat FIT use rechargeable batteries.

TABLE 8.3

Energy Demand for Ear-Worn Wearable Products

Product	Battery Type	Battery Capacity (Wh)	Battery Life (h)	Energy Demand (Wh/h)[a]
General-purpose Wireless Headsets/Headphones				
Apple AirPod	Lithium-ion	0.093	5	0.019
Samsung IconX		0.141	3.8	0.037
Plantronics Backbeat FIT		0.426	7	0.061
Hearing Aids				
Eargo Neo (Hearing Aid)	Lithium-ion	0.074	16	0.005
Eargo Neo (Charger Case)		2.42		N/A
Widex	Zinc 13	0.42	126	0.003
	Zinc 312	0.25	96	0.003

[a] Energy demand is estimated as the ratio of battery capacity to battery life, is given in Wh/h, and is roughly the same as average power consumption in watts.

In the future, ear-based wearables are expected to displace smartphones through the advancement of wireless headphones and headsets. Future versions of these products are expected to provide directions, answer queries, and act as virtual personal assistants, thus facilitating greater hands-free exchanges of information (Gartner 2018). Such expansion has caused ear-worn devices to experience the most rapid growth among wearables despite lagging behind wrist-worn devices in overall sales (Draper 2019). Not surprisingly, the energy demand of ear-based wearables varies depending on their functionality. As is evident from a sample of ear-worn devices in Table 8.3, devices that are general-purpose and designed for frequent communication with smartphones tend to consume the most energy, while those with a specific, custom purpose (e.g., hearing aids) consume far less energy.

8.3.2 On the Wrist

Whether they have a form factor closer to that of a watch or to that of a wristband, wrist-worn devices perform a broad range of functions in a wearable footprint. The term smartwatch is often used to describe a device that is, in some way, dependent on a smartphone for full functionality. These devices include the Apple Watch, smartwatches based on the Wear OS operating system and sold by multiple vendors, and watches based on the Tizen operating system which interacts solely with Samsung Galaxy smartphones. Most smartwatches do a subset of what a smartphone does. They can display notifications, operate applications, manage media playback (e.g., voice, music, video), answer messages using voice dictation, and include a Global Positioning System (GPS) to facilitate location tracking and location-specific notifications. Smartwatches also offer similar functionality to smart wristbands including heart rate, calorie burn, and activity tracking. Some watches sense additional information regarding the body's activity and location in space that is relevant to safety and health. For example, the Apple Watch includes a fall sensor which detects falls, monitors the user for additional movement, and can alert authorities on the user's behalf if no post-fall motion or communication is detected.

Sports watches are different from smartwatches in that they provide functionality that is far less dependent on an accompanying smartphone and is customized to particular activities and behaviors. For example, hiking watches such as the Garmin Fenix 5 (Garmin n.d.) contain a three-axis compass and altimeter, both of which are calibrated automatically using a built in GPS. Atmospheric pressure and ambient temperature are measured with an on-board barometer and thermometer. The watch automatically sets time based on GPS-detected location and can provide directions based on user imported routes. The watch also saves detailed activity information and provides fitness tracking including estimates of oxygen consumption and can be used for running, swimming, skiing, and other sports in addition to hiking. While the durable, robust, abrasion-resistant packages of these watches set them apart from smartwatches and wristbands, their advanced functionality is similar to that of the smartwatch, albeit without the need for a nearby smartphone. Similar application-specific watches are available for diving, flying, swimming, and general-purpose fitness and training. Almost all of these specialty sports watches come at increased cost over smartwatches because they are designed to operate in tough environments.

Smartwatches are similar in form and function to regular watches but also have functionality that requires pairing to a smartphone. Smart wristbands, on the other hand, tend to have less functionality, do not necessarily require frequent pairing to a smartphone, and have a more basic display than a smartwatch. Because of their reduced functionality, smart wristbands also require less power and have much longer battery life. Most wristband batteries, like those in smartwatches, are rechargeable. By far, the most common application for the smart wristband is fitness tracking. 30% of American adults own such a wristband and Fitbit has the highest market share among manufacturers of fitness tracking wristbands (Liu 2019). These wristbands can automatically track exercise based on such activities as running, walking and swimming, monitor heart rate, predict calorie burn, and track female health among other functions. As smart wristbands continue to increase in functionality, their overlap with smartwatches also increases, thus making it increasingly difficult to distinguish between these two types of devices. Although small in number, some smart wristbands have also been successfully demonstrated for functions other than fitness tracking. For example, a touch-sensitive electronic wristband can be attached to a smartwatch to facilitate keyboard style interaction with the smartwatch (Funk et al. 2014) that is not possible within the limited display area of the watch alone. Sunu (Sunu Inc n.d.) makes a wristband that uses sonar to detect nearby obstacles for the visually impaired and provide feedback through vibratory patterns

TABLE 8.4

Energy Demand for Wrist-Worn Wearable Products

Product	Battery Type	Battery Capacity (Wh)	Battery Life (h)	Energy Demand (Wh/h)[a]
General-Purpose Smartwatch				
Apple Watch 4	Lithium-ion	1.11	18	0.062
Fitbit Versa		0.54	96	0.006
Samsung Galaxy (46 mm)		1.75	48	0.036
Samsung Galaxy (42 mm)		1.00	48	0.021
Fitness Trackers				
Fitbit Charge 2	Lithium-polymer	0.22	120	0.002
Huawei Honor Band 3		0.37	15	0.025
Samsung Gear Fit2 Pro	Lithium-ion	0.74	72	0.010
Sports Watch				
Garmin Fenix 5 (Basic Fitness)	Lithium-polymer	0.85	336	0.003
Garmin Fenix 5 (GPS)			75	0.011
Garmin Fenix 5 (GPS+)[b]			24	0.035

[a] Energy demand is estimated as the ratio of battery capacity to battery life, is given in Wh/h, and is roughly the same as average power consumption in watts.

[b] GPS + includes precise positioning including dead reckoning using gyro sensors.

to the user. These two wristband devices suggest a wide breadth of potential applications that have yet to be fully explored for wrist-worn wearables.

As with ear-worn wearables, those devices made for general-purpose use and frequent communication with smartphones tend to consume the most energy. The energy demand estimates in Table 8.4 confirm this to be the case with smartwatches, on average, consuming more power and having higher energy demand than other wrist-worn devices.

8.3.3 On the Head

Approximately 80 million head-mounted displays were sold worldwide in 2018 (Gartner 2018). These head-worn wearables can be loosely classified into classical, augmented reality, and virtual reality designs. Classical designs are straightforward and display a standard image or video onto a screen in a head-worn unit designed to provide hands-free convenience in passively watching visual content. Augmented reality, head-mounted displays overlay an image or images onto what is front of the user, providing the impression that the digital overlay is interacting with or woven into the present reality. And, virtual reality creates a whole different reality for the user and, as such, promotes interaction with virtual worlds rather than reality. The subtle difference between classical and virtual reality head-mounted displays is that classical units typically do not involve active user participation while virtual reality units do. Classical units are increasingly displaced by augmented and virtual reality units.

Historically, helmet-mounted displays have dominated technology that is worn on the head (excluding ear-worn technology), while wearable smart glasses are becoming more popular due to their light weight and greater convenience of use. Defense applications dominate demand for helmet-mounted and other head-mounted displays followed in decreasing order by applications in industry, healthcare, and consumer electronics. However, as cost continues to decline, consumer electronic demand is expected to drive future growth in head-worn wearables (Ergürel 2016).

In defense applications, head-mounted displays are used to display tactical information (e.g., maps, thermal imaging) that is overlaid onto actual scenes (i.e., augmented reality) and are typically integrated with other instruments and devices into a rugged and expensive package. In industry, head-mounted displays are used to support 3D visualization in computer-aided design, enable flight, vehicle, and other simulations that are impractical to replicate in the real world, and support training for medical practitioners, soldiers, welders, and more. In medicine, these wearable devices augment surgeon capabilities in orthopedic, laparoscopic, cancer resection, endoscopic, and other procedures and support patients with compromised vision, driving ability, and other limitations (Iqbal et al. 2016). The energy demand for these head-worn devices tends to increase with the quality and resolution of the display (Table 8.5).

TABLE 8.5

Energy Demand for Head-Worn Wearable Products

Product	Application	Battery Capacity (Wh)	Battery Life (h)[a]	Energy Demand (Wh/h)[b]
Virtual Reality (VR) Head-Worn Displays				
HTC Vive Wireless (VR)	General purpose	37.2	2.5	14.9
Oculus Go (VR)	Gaming	9.36	2	4.68
Augmented Reality (AR) Head-Worn Displays				
Microsoft HoloLens (AR)	General purpose	61.1	2	30.5
Solos Smart Glasses (AR)	Cycling	1.48	5	0.296
ODG R-7H (AR)	Heavy industry; mining	4.81	5	0.962
Vuzix M300 (AR)	Industry; manufacturing	0.592	2	0.296

[a] Hours of typical use before recharging is required.
[b] Energy demand is estimated as the ratio of battery capacity to battery lifetime, is given in Wh/h, and is roughly the same as average power consumption in watts.

8.3.4 Smart Clothing

Although sales of smart clothing continue to lag behind devices worn on the head, wrist, or ear, smart clothing nevertheless contributes to the growth of wearable technology. In fact, some surmise that the reason that smart clothing has not kept pace with other wearables is that many consumers are not aware of the types of smart clothing that are commercially available. Recent upticks in Google searches on smart clothing may indicate growing awareness and this is likely to stimulate future sales (Hanuska et al. n.d.). Smart clothing is used in the military, entertainment, sports, healthcare, and other market sectors to protect, augment, and inform. Such clothing can be *passive* (i.e., sense but not respond), *active* (i.e., sense and respond), or *very active* (i.e., sense, respond, and adapt).

For example, the SmartShirt is a *passive* technology used in healthcare which connects to and integrates information from a wide variety of body sensors and gathers heart rate, electrocardiogram (ECG), respiration, blood pressure, and other vital signs. The SmartShirt is often used among first responders and other emergency personnel to reduce the risk of injury and death. Electro-optical fibers embedded in the shirt fabric collect body data and send that information wirelessly to a doctor or other professional for assessment and decision making. The goal of the SmartShirt is to improve occupational safety among first responders in a wide range of emergency situations (Cho, Lee, and Cho 2010).

Another examples of a *passive* smart clothing product in healthcare is the LifeShirt®. This smart shirt monitors respiration by measuring movement of the chest and abdominal wall using a technique called respiratory inductance plethysmography. Bands containing piezoresistors are sewn into a Lycra vest to monitor the movement of the chest and to estimate lung volume from respiratory movements. The vest also uses ECG electrodes to monitor heart activity. Data collected from both piezoresistor and ECG sensors is stored in a module carried in the pocket or belt and transferred via the internet to a secure location for interpretation by medical personnel (Cho, Lee, and Cho 2010).

Athos (Athos n.d.) offers both shorts and shirts to the athlete for *passive*, real-time tracking of muscle activity, heart rate, and calorie burn both during active and rest time. Sensor data is collected and sent to the user's smartphone which can, if desired, be checked almost in real-time. Real-time data allows for timely adjustments in movement and behavior that work to improve performance and avoid possible injury. The compression fabric is sweat-wicking and contours closely to the body to ensure more accurate measurements.

The Musical Jacket is an *active* smart clothing system which consists of a MIDI keyboard embroidered into the jacket fabric using conductive thread made of stainless steel. Body movement is detected capacitively when part of the thread makes contact with the body. Multiple touches are possible and the resulting signal or signals are transferred into a MIDI synthesizer and microcontroller that interprets the signals and conveys the musical signal back out to speakers in the pocket of the jacket. The entire system is stand-alone and operates on batteries (Cho, Lee, and Cho 2010).

Still other wearable technology is not sensor-based at all. For example, Illumio (Illumio n.d.) makes a machine washable jacket with fully integrated LEDs including rechargeable battery that can be seen hundreds of feet away to protect the user in dimly lit or dark environments. While not containing any sensors, the Illumio jacket has one thing in common with a vast majority of wearables on the market—it consumes power—and plenty of it.

Finally, as a testament to what a truly comprehensive, *very active* wearable might look like, the US military initiated development of the TALOS exosuit. In collaboration with US corporations, universities, national laboratories, and government agencies, this full integrated, technology-loaded smart clothing system is focused on monitoring and protecting American soldiers in combat. The suit was designed to sense body temperature, heart rate, and hydration levels and use this data to assess and respond to wounds and injury. TALOS was intended to be bullet proof across the full body and to remain flexible, stiffening only when needed to provide additional protection or joint support. In addition to providing multiple sensor cues to soldiers to prevent injury and support performance, the suit was intended to harvest energy and reapply it to movement where the solider needed it most. All of these functions were to be provided within a lightweight 400 pound footprint including a means of providing 12 kW of power for 12 hours at a time (Hoarn 2014; Scataglini, Andreoni, and Gallant 2015). Unfortunately, by 2019, the TALOS project was cancelled due to insurmountable technical challenges and excessive power consumption (Keller 2019).

8.3.5 Other Wearables

Although commercial wearables favor placement on the head, ears, or wrist, electronic devices worn on other places on the body are proliferating as well. For example, the AIO sleeve (Komodo Technologies n.d.) worn on the arm provides increased accuracy regarding heart rate and heart beat than wearable technology that is wrist-worn. Most wrist-worn devices offer information regarding heart rate through optical means. A light is projected into the wrist and is absorbed more when the heart beats (i.e., higher blood flow) and less when the heart is between beats. While convenient and inexpensive, this method of heart rate detection is less accurate than conventional electrical methods using an electrocardiogram (ECG). To provide increased accuracy, the AIO sleeve tracks the electrical behavior of the heart in a manner similar to an ECG. The AIO sleeve uses this data to electrically monitor workout intensity, sleep, and other daily activities with greater accuracy than most wrist-worn fitness trackers.

For individuals who find a wristband or sleeve too cumbersome, wearable rings are an alternative for fitness tracking. Motiv (Motiv n.d.) tracks heart rate during sleep, workouts, and other activities with sufficient on-board memory to eliminate the need to pair the ring with a smartphone during use. The Ouraring (Oura n.d.) provides additional functionality with sensors that optically measure heart rate and pulse and in combination with accelerometers, gyroscopes, and pressure sensors, monitors both sleep quantity and sleep quality. Other alternatives for fitness tracking include the Spire Health tag (Spire Health n.d.) which can be placed almost anywhere on the body that involves clothing and the Fitbit clip (Fitbit n.d.) that can be attached to waistband, belt, bra, or any piece of clothing that remains in close and consistent contact with the body.

Wearable technology certainly does not end at the wrist or hands but often travels downward, landing on the foot. Wearable technology (Runscribe n.d.) communicates information to the runner about cadence, contact times (between foot and ground), pronation (side-to-side motion of the foot), velocity, shock, and braking force to a smartphone and associated app—all via two clips mounted to the top of the shoes. Fitness socks (Sensoria n.d.) go a step further by integrating pressure sensors into the textile of the sock itself and providing real-time audio cues when a running style that is vulnerable to injury is detected. For a more accurate portrayal of what the feet experience, pressure-sensitive insoles (Arion n.d.) measure interactions between the feet and ground directly. Shoe-worn wearables don't stop with fitness and sport, however. Wearable technologies in the shoe have also been used to support a greater gaming experience. For example, Bcon (Bcon n.d.) makes a device that is mounted on a sneaker or directly onto the foot to supplement a mouse and keyboarding during video game play. The foot-worn system provides 24 additional keys and means to control the gaming experience, enough additional control to give a leg (or foot) up to those who are serious about gaming.

Shoe or foot-worn wearables are also useful for monitoring gait during walking, running, and similar activities. But, the foot is not the only place where wearables can be mounted to provide useful gait information. In addition to being placed on the feet, wearable accelerometers are also available for placement on the lower back, ankle, waist, around the neck, and on the chest for monitoring basic spatiotemporal gait parameters including stride length, cadence, and speed as well as more complex kinematics associated with both normal and abnormal gait. The portability and accessibility of wearable devices makes real-world gait monitoring possible but also opens up a tremendous need for greater databases of sensor information associated with specific gait abnormities, gait types, and demographic-specific information (Benson et al. 2018).

Posture is an important part of good musculoskeletal health and wearables to monitor posture at all levels from head to toe are also available commercially. DorsaVi (Dorsavi n.d.) makes a device containing both motion and muscle activity sensors to monitor body movement for a wide range of applications in workplace safety, sports, and physical therapy. The medical grade sensors are paired with external video collected from cameras to evaluate potential risk of injury and optimize movement for greater overall health, safety, or rehabilitation. The Upright Go (Upright n.d.) is a little more consumer friendly and less expensive than DorsaVi products; the wearable sensor module mounts to the middle of the upper back to provide real-time posture information via pairing to a smartphone app. LumoLift (Lumo n.d.) monitors the curvature at the top of the spine as well as the positions of the shoulders, chest, and upper back and provides vibrational feedback to indicate poor posture for real-time posture adjustment. Other wearables go a step further, providing electronic muscle support to offset weakness in the torso, hips, and legs for greater strength in getting up, sitting down, or similar movements (Takahashi 2017).

8.4 Overall Energy Demand

While the future of humankind may be facing a shortage in drinking water, it is certainly not facing a shortage of portable and wearable devices that individuals are willing to haul around on their person. Powering all of these devices and doing so sustainably and on-demand is a challenge. Estimating the demand is no less challenging. While battery life for wearables and portables is available for most consumer devices because it is of direct interest to the users, power consumption and battery capacity are much more difficult to determine. The newer or more novel the device, the harder it is to find sufficient information to determine energy demand. But, it is these cutting edge devices that often define the trends of the future and in turn, determine the energy demand incurred by those trends.

Multiple strategies can be used to close in on the ball park of the target energy (and power) that wearable solar cells need to provide to wearables and portables. One approach is to simply target the capacity of commercially available, portable power banks. These power banks are available in a range of capacities, from 3,350 to 28,600 mAh (Table 8.6). The energy that the power bank can realistically supply to one or more portable devices is then:

$$\text{Energy (in Wh)} = 0.001 \times \text{Capacity (in mAh)} \times \eta \times V_{cell} \qquad (8.1)$$

where the cell voltage (V_{cell}) can be assumed to represent typical lithium-based batteries at between 3.6 and 3.8 V, and the efficiency of the boost (or buck) converter (η) represents the power lost in having to convert the battery cell output voltage to the voltage required by the device being charged. Many devices are set up to charge on a 5-V USB standard, but laptop computers and other high-energy-demand devices often require even higher voltages, up to 20 V. Boost conversion efficiencies are typically between 90% and 98%, while buck conversion efficiencies trend lower.

TABLE 8.6

Energy Capacity of Commercially Available Power Banks

Power Bank	Battery Capacity (mAh)	Energy Capacity (Wh)[a]	Phones	USB Compatible Wearables	Tablets	Laptops
Anker PowerCore 10000	10,000	35.2	Yes	Yes	No	No
Anker PowerCore+ 26800PD	28,600	100.5	Yes	Yes	Yes	Yes
Belkin Pocket Power 5k	5,000	17.6	Yes	Yes	Yes	No
iMuto Portable Charger X6	30,000	105.5	Yes	Yes	Yes	Yes
iQunix MiniPower	3,350	11.8	Yes	Yes	No	No

[a] Assumes a 3.7-V battery cell voltage and 95% boost converter efficiency.

Striving to duplicate the energy capacity of commercially available power banks using wearable solar cell systems requires making some assumptions about how often the capacity is required. Does the user of the power bank need a recharge over the course of a business day (8 hours)? Daylight hours (12 hours)? A full day (24 hours)? Multiple days? How often does the typical tech-savvy, tech-laden individual recharge a portable power bank? Knowing this information is important to all power bank designs, whether energy is ultimately harnessed from what comes out of a wall outlet, from the sun, or from some other source of energy. Unfortunately, it is difficult to predict individual energy demand patterns, particularly those of the future, which will be driven by even more tech-savvy individuals.

Another approach to estimating total energy demand incurred by portables and wearables is more compatible with available information regarding projected growth rates in wearable and portable device market sectors. Based on existing information such as that provided in Tables 8.1 through 8.5, it is possible to estimate energy demand by device category beginning with two major classes of portables:

- High energy demand (laptop, e-readers, and tablet computers): The broad functionality, high-speed, and high-quality displays expected of these devices will keep energy demand high as battery technology improvements are offset by increased performance and functionality.

- Moderate energy demand (mobile phones): Smartphones already dominate mobile phone markets and are likely to continue taking over traditional flip phones in the future. A survey of several smartphone products from several popular smartphone manufacturers (Table 8.1) provides an estimated range between 6.35 and 18.5 Wh of energy demand over a typical 24-hour day or between 0.26 and 0.77 Wh/h of use.

While smartphone use is expected to continue to displace laptop and tablet computer use, the functionality of the smartphone will have to increase to keep pace with consumer expectations. Therefore, for the foreseeable future in general-purpose computing, it is expected that more individuals will be carrying around smartphones with greater energy demand while fewer will be carrying around laptop computers.

In addition to portables, wearables also have the potential to place a significant energy demand on portable power sources:

- Very high energy demand (wearables with high-quality, heavy-usage displays): A sample of commercially available devices indicates that virtual reality displays have high energy demand that is sometimes even greater than that of laptop and tablet computers.

- Moderate energy demand (general-purpose wearables with medium-quality, moderate-usage displays): The smartwatches and wireless headsets exemplify the demand for general-purpose wearables. A survey of popular products in this category estimates energy demand between 0.006 and 0.062 Wh/h of usage. However, these estimates do not account for the increased energy required of the smartphone to communicate with these devices. Holistically, the real energy demand of these general-purpose wearables, both due to the need for a display and due to frequent interaction with a smartphone, is likely to be higher than appears at first estimate.

- Low energy demand (limited functionality wearables with moderate use displays): These wearables do not require pairing with a cell phone for most or all of their functionality and often have minimalist displays, which limits their energy demand to the low end of those surveyed (between 0.002 and 0.035 Wh/h of usage). Unlike the general-purpose wearables, these devices only interact with smartphones or other computing devices intermittently, which keeps their energy demand low.

- Very low energy demand (wearables with narrow functionality and no display): These wearables are typically designed for a very specific purpose (e.g., standalone MP3 players, hearing aids), have minimal or no display/screen, and have infrequent communication with other devices including smartphones. Energy demand is typically less than 5 mWh/h (0.005 Wh/h).

Using these guidelines, it is possible to estimate the total amount of mobile energy demand associated with a particular suite of wearable devices (Table 8.7). These estimates are a good first step toward choosing a solar cell or other energy harvesting technology that can adequately serve such demand.

TABLE 8.7

Energy Demand Guideline per Device Type

Type of Device	Energy Demand (Wh/h)[a]	
Virtual reality, general-purpose wearables	Very high	4.7–30.5
Laptop computers	High	1.3–6.3
Tablet computers	High	2.1–7.5
E-Readers	Moderate	0.25
Smartphones	Moderate	0.26–0.77
Augmented reality, application-specific wearables	Moderate	0.3–1.0
Fitness trackers & similar minimal display devices	Low	0.002–0.025
Smartwatches	Low	0.006–0.062
Sports watches	Low	0.003–0.035
Hearing aids and similar application-specific devices	Very low	0.003–0.005

[a] Energy demand is roughly the same as average power consumption in watts.

8.5 Summary

While the total energy required to power an individual's sum total of wearables and portables may seem small compared to residential power demand, the surface area over which an individual can wear solar panels is also quite a bit smaller than the typical rooftop. Further adding to the challenge, the batteries, battery holders, and replacement and recharging requirements (both capacity and speed) for wearable devices are often as many and varied as the devices themselves. In the future, the consumer is likely to be faced with an ever-expanding and dizzying array of wearable gadgets to choose from, and more and more wearable technology will be required in military, in industry, and even in healthcare to ensure safer, healthier, and more reliable operations. With this inevitable upswing in wearables will emerge a greater interest for alternatives to the standalone battery-based power bank for powering these devices. Given the abundance of light both from natural sunlight and artificial light sources in an individual's everyday environment, the prospect of meeting this energy demand with wearable solar cells and systems is inviting. Wearable solar cells remain an attractive option despite the complications introduced by a highly dynamic, mobile platform (e.g., the human body) on which these solar cell systems must operate.

References

Apple. n.d. "AirPods." Apple. Accessed July 3, 2019. https://www.apple.com/airpods/.
Arion. n.d. "ARION—Transform Your Running Technique." Accessed July 15, 2019. https://www.arion.run/.

Athos. n.d. "Athos Training System." Athos. Accessed July 5, 2019. https://shop.livea-thos.com.

Bcon. n.d. "World's First Gaming Wearable." Accessed July 15, 2019. https://bcon.zone/.

Benson, Lauren C., Christian A. Clermont, Eva Bošnjak, and Reed Ferber. 2018. "The Use of Wearable Devices for Walking and Running Gait Analysis Outside of the Lab: A Systematic Review." *Gait & Posture* 63 (June): 124–138. https://doi.org/10.1016/j.gaitpost.2018.04.047.

Cho, Gilsoo, Seungsin Lee, and Jayoung Cho. 2010. "Review and Reappraisal of Smart Clothing." In *Smart Clothing: Technology and Applications*. Boca Raton, Florida: CRC Press.

Dorsavi. n.d. "DorsaVi Global—Wearable Sensor Technology & Movement Assessment." Accessed July 15, 2019. https://www.dorsavi.com/.

Draper, Sam. 2019. "IDC: Wrist-Worn and Ear-Worn Wearables Lead the Market, Fueled by a Strong Growth Trajectory." *Wearable Technologies* (blog). June 4, 2019. https://www.wearable-technologies.com/2019/06/idc-wrist-worn-and-ear-worn-wearables-lead-the-market-fueled-by-a-strong-growth-trajectory/.

Ergürel, Deniz. 2016. "Lower Costs Will Drive the Head Mounted Display Market." Haptical. September 14, 2016. https://haptic.al/head-mounted-display-report-technavio-2016-a5ee5fe36baf.

Fitbit. n.d. "Fitbit Official Site for Activity Trackers & More." Accessed July 15, 2019. https://www.fitbit.com/home.

Franklin, Neil. 2017. "More People Will Have Smartphones than Running Water or Bank Accounts by 2021, Claims Report." Workplace Insight. February 14, 2017. https://workplaceinsight.net/people-will-smartphones-running-water-bank-accounts-2021-claims-report/.

Funk, Markus, Alireza Sahami, Niels Henze, and Albrecht Schmidt. 2014. "Using a Touch-Sensitive Wristband for Text Entry on Smart Watches." In *CHI '14 Extended Abstracts on Human Factors in Computing Systems*, 2305–2310. CHI EA '14. New York, NY: ACM. https://doi.org/10.1145/2559206.2581143.

Garmin. n.d. "Garmin Fēnix® 5S Plus | Multisport GPS Watch." Garmin. Accessed July 5, 2019. https://buy.garmin.com/en-US/US/p/603201.

Gartner. 2018. "Gartner Says Worldwide Wearable Device Sales to Grow 26 Percent in 2019." Gartner. https://www.gartner.com/en/newsroom/press-releases/2018-11-29-gartner-says-worldwide-wearable-device-sales-to-grow-.

Grand View Research. 2016. "Wearable Technology Market Size, Share." https://www.grandviewresearch.com/industry-analysis/wearable-technology-market.

Grush, Andrew. 2019. "Android Smartphones with the Best Battery Life." Android Authority. https://www.androidauthority.com/best-android-phone-battery-life-2-755699/.

Hanuska, Alex, Bharath Chandramohan, Laura Bellamy, Pauline Burke, Rajiv Ramanathan, and Vijay Balakrishnan. n.d. "Smart Clothing Market Analysis." https://scet.berkeley.edu/wp-content/uploads/Smart-Clothing-Market-Analysis-Report.pdf.

Hoarn, Steven. 2014. "Tactical Assault Light Operator Suit (TALOS) Doesn't Lack Ambition or Interested Industry." Defense Media Network. May 21, 2014. https://www.defensemedianetwork.com/stories/diverse-range-of-companies-work-on-the-tactical-assault-light-operator-suit-talos/.

Illumio. n.d. "Higher Visibility Illumio Jackets." Accessed July 15, 2019. https://myil-lumio.com/.

Iqbal, Mohammed H., Abdullatif Aydin, Oliver Brunckhorst, Prokar Dasgupta, and Kamran Ahmed. 2016. "A Review of Wearable Technology in Medicine." *Journal of the Royal Society of Medicine* 109 (10): 372–380.

Keller, Jared. 2019. "SOCOM's Iron Man Suit Is Officially Dead." Task & Purpose. https://taskandpurpose.com/talos-iron-man-suit-dead.

Komodo Technologies. n.d. "AIO Smart Sleeve—ECG Wearable & Heart Rate Variability Monitor." Accessed July 15, 2019. http://komodotec.com/.

Liu, Shanhong. 2019. "Fitness & Activity Tracker." May 22, 2019. https://www.statista.com/topics/4393/fitness-and-activity-tracker/.

Lumo. n.d. "Lumo Bodytech." Accessed July 15, 2019. https://support.lumobodytech.com/hc/en-us.

Motiv. n.d. "Motiv Ring." Motiv. Accessed July 15, 2019. https://mymotiv.com/.

Osborne, Joe. 2017. "Microsoft Surface Pro (2017) Review." TechRadar. https://www.techradar.com/reviews/microsoft-surface-pro/2.

Oura. n.d. "Oura Ring: The Most Accurate Sleep and Activity Tracker." Accessed July 15, 2019. https://ouraring.com/.

Pew Research Center. 2015. "U.S. Technology Device Ownership: 2015." http://www.pewinternet.org/2015/10/29/technology-device-ownership-2015/.

Pew Research Center. 2019a. "Smartphone Ownership Is Growing Rapidly Around the World, but Not Always Equally." Pew Research Center's Global Attitudes, Project (blog). February 5, 2019. https://www.pewresearch.org/global/2019/02/05/smartphone-ownership-is-growing-rapidly-around-the-world-but-not-always-equally/.

Pew Research Center. 2019b. "Use of Smartphones and Social Media Is Common across Most Emerging Economies." March 7, 2019. https://www.pewinternet.org/2019/03/07/use-of-smartphones-and-social-media-is-common-across-most-emerging-economies/.

Plantronics. n.d. "BackBeat FIT 2100, Wireless Sport Headphones." Accessed July 3, 2019. https://www.plantronics.com/us/en/product/backbeat-fit-2100.

Reisinger, Don. 2019. "Best MP3 Players 2019." Tom's Guide. May 28, 2019. https://www.tomsguide.com/us/best-mp3-players,review-6107.html.

Rioja, Alejandro. 2019. "17 Best Smartphones with Largest Battery Capacity." Flux Chargers. https://www.fluxchargers.com/blogs/flux-blog/best-smartphones-largest-battery-capacity-life.

Runscribe. n.d. "RunScribe Gait Analysis Platform." Accessed July 15, 2019. https://runscribe.com/.

Samsung. n.d. "Gear IconX." The Official Samsung Galaxy Site. Accessed July 3, 2019. http://www.samsung.com/global/galaxy/gear-iconx/.

Scataglini, Sofia, Giuseppe Andreoni, and Johan Gallant. 2015. "A Review of Smart Clothing in Military." In *Proceedings of the 2015 Workshop on Wearable Systems and Applications*, 53–54. WearSys '15. New York, NY, USA: ACM. https://doi.org/10.1145/2753509.2753520.

Sensoria. n.d. "Sensoria Artificial Intelligence Sportswear." Accessed July 15, 2019. https://www.sensoriafitness.com/.

Spire Health. n.d. "Spire Health: Clinical-Grade Health Monitoring and Insights." Accessed July 15, 2019. https://spirehealth.com/.

Statista. 2018. "Share of Households with a Computer Worldwide 2005–2018." Statista. https://www.statista.com/statistics/748551/worldwide-households-with-computer/.

Statista. 2019. "Tablet Ownership among U.S. Adults 2010–2019." Statista. https://www.statista.com/statistics/756045/tablet-owners-among-us-adults/.

Statista. n.d. "Global E-Book Reader Shipments 2008–2016." Statista. Accessed July 21, 2019. https://www.statista.com/statistics/272740/global-shipments-of-e-book-readers/.

Strom, Karl. 2018. "Will Hearing Aid Sales Top 4 Million Units in 2018?" Hearing Review. http://www.hearingreview.com/2018/11/will-hearing-aid-sales-top-4-million-units-2018/.

Sunu Inc. n.d. "Sunu Band." Accessed July 15, 2019. https://www.sunu.com/en/index.html.

Tabari, Rammi. 2019. "Laptops with Best Battery Life 2019—Longest Lasting Laptop Batteries." https://www.laptopmag.com/articles/all-day-strong-longest-lasting-notebooks.

Takahashi, Dean. 2017. "Superflex Unveils Powered Super Clothing Created with Designer Yves Béhar." *VentureBeat* (blog). January 11, 2017. https://venturebeat.com/2017/01/11/superflex-unveils-powered-super-clothing-created-by-designer-yves-behar/.

The Hearing Review. 2018. "EHIMA Data Shows 5.7% Increase in Global Hearing Aid Sales." Hearing Review. http://www.hearingreview.com/2018/05/ehima-data-shows-increase-global-hearing-aid-sales/.

Upright. n.d. "Upright Posture Training Device—Everyday Posture Coaching." Accessed July 15, 2019. https://www.uprightpose.com/.

Veloxity. 2017. "Cell Phone Battery Statistics 2015 2016 2017." https://veloxity.us/2015-phone-battery-statistics/.

9

Wearable Solar Systems

Designing a wearable solar cell system invokes a very different set of considerations than traditional solar cell system designs. While both types of systems begin with the selection of a photovoltaic (PV) material, a traditional solar cell system design is driven by efficiency gains within the material itself. The primary design goal of these systems is often to simply generate the most power in the smallest area possible. In wearable solar cell systems, however, energy demand is much lower, and even with the reduced energy input that comes from exposure to a mixture of natural sunlight and artificial light, PV cell efficiency may play less of a singular role in overall design considerations. Instead, other considerations such as conformity to irregular topologies, lack of toxicity, materials that operate well under low light energy, and electronics that tolerate wide swings in input conditions become more important.

9.1 Basic Performance

The basic performance of a PV material is the first step in choosing the type of PV cell which will serve as the foundation of a successful wearable solar cell system. To this end, the latest and greatest efficiencies of basic bilayer (single junction) solar cells in response to AM1.5 testing conditions (i.e., sunlight) are summarized in Table 9.1 for all the major players in first-, second-, and third-generation solar cell technologies. While efficiency is not the beginning and end of the story in designing a wearable solar cell system, it is a critical piece of the design puzzle.

PV materials like monocrystalline silicon are mature and have experienced little improvement over the last decade. Other materials such as quantum dot PV and perovskite materials are less mature and have experienced more rapid improvements over this same time period. Both efficiency and improvements in efficiency can be important factors in choosing a PV material. Larger improvements are suggestive of the PV material's potential for the future while present efficiency is a gateway issue that, if too low, can eliminate the PV material from consideration altogether. Because the total energy demand of wearable systems is almost certain to increase, the future potential of wearable PV is as relevant as its present capability for generating electrical energy from light energy.

TABLE 9.1

Basic Performance Metrics in Response to AM1.5 (Terrestrial Sunlight)

PV Technology	Efficiency[a] (%)	Improvement[b] (%)	V_{oc}[c]	Fill Factor[c]
Monocrystalline Si	26.1	1.9	0.70	0.83
Polycrystalline Si	22.3	2.3	0.68	0.80
GaAs (thin film)	29.1	3.0	1.12	0.86
Amorphous Si (thin film)	14.0	2.0	0.52	0.73
CIGS (thin film)	22.9	3.5	0.72	0.79
CdTe (thin film)	22.1	5.7	0.85	0.79
Organic	15.6	9.4	0.80	0.72
Dye-sensitized	11.9	1.0	0.85	0.72
Perovskite	24.2	NA[d]	1.08	0.74
Quantum dot	16.6	13.6	0.65	0.72

[a] Best research cell efficiency as of 2019 (NREL 2019).
[b] Increase in best demonstrated efficiency between 2009 and 2019 (NREL 2019).
[c] Open circuit voltage and fill factor gathered from 2016 review of PV cells (Polman et al. 2016).
[d] Research cell efficiencies were not reported for perovskite PV cells until 2013.

Since wearable solar cell systems will be used not only outdoors where they will be exposed to the energy of the sun but indoors where they will be exposed to artificial lighting sources, efficiencies under both sets of conditions must also be considered in the selection of a PV material. Because of the different spectral responses among PV materials and unique spectrum of different light sources, efficiencies can vary dramatically for different combinations of PV material and light source. Traditional solar cell systems are almost universally exposed to sunlight, so it is not surprising that PV cells now demonstrate their best efficiencies in response to sunlight. Table 9.2 summarizes efficiencies for bilayer (single junction) PV cell designs in response to sunlight and various forms of artificial light. More complex system designs like tandem cells and solar concentrators can provide even greater efficiencies than those summarized in Table 9.2.

However, concentrator designs are entirely impractical for wearable systems because they require direct light to be effective and tracking devices to maximize access to direct light. Although they are more expensive than single junction cells, tandem cells are more practical than concentrators for wearable systems, but remain fundamentally limited in efficiency and performance by the underlying PV material. While tandem cells can add to the amount of light energy that is collected by a PV cell, they do not change the underlying collection efficiencies of the PV cell once light is converted into excitons or electron-hole pairs. Most tandem cells are designed to collect more of the wavelength spectrum (i.e., more colors of light) of sunlight. Tunable PV cells, such as those made possible by adjusting the size of quantum dots or

TABLE 9.2

Solar Cell Efficiency under Artificial Light Sources

PV Technology	Relative Efficiency[a]				
	Sunlight[b]	Fluorescent[c]	Incandescent[d]	LED[e]	Metal Halide
Monocrystalline Si	1.00	0.34	3.40	0.38	0.48
Polycrystalline Si	0.85	0.29	2.90	0.32	0.42
GaAs (thin film)	1.11	0.52	2.23	0.58	0.72
Amorphous Si (thin film)	0.54	0.39	0.67	0.40	0.47
CIGS (thin film)	0.88	0.30	2.65	0.39	0.44
CdTe (thin film)	0.85	0.42	1.52	0.49	0.59
Organic (OPV)[f]	0.60	0.43	0.60	0.45	0.51
Dye-sensitized (DSC)	0.46	0.24	0.76	0.29	0.34
Perovskite[g]	0.93				
Quantum dot[g]	0.64				

Source: Minnaert and Veelaert (2014) and NREL (2019).

[a] Efficiencies are calculated relative to that of monocrystalline silicon (under AM1.5 illumination conditions).

[b] Based on best research cell efficiencies within each PV technology.

[c] Cool white fluorescent lamp with a correlated color temperature (CCT) of 4,230 K (typical office illumination).

[d] Represented by a black body at a temperature of 2,856 K.

[e] Warm LED (cool LED efficiencies are slightly different).

[f] Efficiencies are calculated for an organic PV cell with a DTDCTP:C70 active layer.

[g] Efficiencies for indoor light sources vary depending on (tunable) bandgap and material.

the composition of perovskite cells, are desirable in traditional solar cell systems to facilitate collecting more of the wavelength spectrum of incoming sunlight. In wearable solar cell systems, however, tunable PV cells may be attractive for an entirely different reason. In these systems, it may be possible to tune PV cells in tandem structures or microarrays of PV cells so that each cell is optimally tuned to a particular light source. Unlike single cell designs whose efficiency has been tuned to sunlight and drops off for other light sources, such designs could convert light to electricity with similar efficiencies, regardless of the type of ambient light source and corresponding wavelength spectrum.

Whether in a traditional single junction design or in a novel tunable tandem cell system that responds well to multiple types of ambient light sources, efficiency is only one piece of the basic performance puzzle. Like efficiency, the open circuit voltage (Table 9.1) of the cell is highly dependent on the underlying PV material used to make the cell. While open circuit voltage does not directly impact the efficiency by which light is converted to electricity, it does influence PV array design. In general, the smaller the open circuit voltage, the more cells are required in a string of PV cells and the more overall PV cells are required in a PV array. For example, both monocrystalline and polysilicon bilayer (single junction) PV cells have open circuit voltages of approximately 0.7 V while GaAs junctions have an open circuit voltage of approximately 1.1 V. This means that a maximum of only 4 or 5 GaAs PV cells need be connected in series to support a 3.7-V lithium battery while between 6 and 8 silicon PV cells may be required per string to provide sufficient voltage to the same battery. In traditional installations which require that many more PV cells be connected in series, differences in open circuit or cell voltage can lead to a major difference in the total number of PV cells required to make a complete system. However, with the assistance of highly efficient boost conversion and the reduced step-up requirements associated with wearable solar cell systems, a string of more than two or three cells in series, regardless of PV material, can be sufficient to provide adequate voltage for most wearables and some portables like smartphones and MP3 players. More PV cells in series may be required to support charging laptop computers because of their higher operating voltages, which approach 20 V.

The fill factor of a PV cell is another basic performance metric that contributes to the goodness of a solar cell system, whether traditional, wearable, or otherwise. The fill factor indicates how well electrical energy is extracted from the cell once light energy is converted to electrical energy from the PV effect. The fill factor reflects losses due to recombination of electrons and holes as well as other traps and defects that capture energized electrons and holes that might otherwise be collected as useful electrical current (electricity). Fill factor is defined as the ratio between the maximum power extracted from a real PV cell and the power that could be extracted from an ideal PV cell (i.e., the product of open circuit voltage and short circuit current). Higher fill factors indicate a PV cell which loses less electrical energy while higher efficiencies reflect a PV cell which loses less light energy. Not surprisingly given its maturity among PV technologies, crystalline silicon outperforms most other PV cell technologies in both fill factor and basic efficiency (Table 9.1).

9.2 Flexibility, Cost, Toxicity, and Stability

Selecting the best PV technology for a wearable solar cell system involves more than considering efficiency, open circuit voltage, and fill factor. For example, one PV material may offer high efficiency but a less efficient PV material may offer greater surface area coverage and thereby produce as much or more useful electricity than a more efficient but rigid PV material. Thus, to generate sufficient power, a wearable solar cell system must either consist of highly efficient, rigid PV cells (e.g., monocrystalline or polycrystalline silicon) or sufficiently flexible, alternative PV cells to cover a larger and more irregular topology and make up for reduced efficiency through greater surface area coverage. Mechanical flexibility is an advantage that less efficient PV cells such as those made with organic and perovskite materials possess, compared to most first-generation, silicon-based solar cells. Organic (OPVs) and perovskite solar cells (PSCs) are not only inherently flexible but many are compatible with low-cost inkjet printing and similar techniques. These manufacturing methods can be used to imprint PV cells directly onto clothing, backpacks, shoes, and other wearable products.

Lower costs are also possible with other second- and third-generation solar cell technologies including many dye-sensitized solar cells (DSCs), quantum dot solar cells and thin film semiconductor PV technologies made with CdTe, CIGS, and amorphous silicon. Thin film GaAs is also flexible as a result of its thin geometries. However, GaAs is often too expensive for consumer wearables. CIGS, while low cost, is not a good candidate for many wearable solar cell applications because it performs poorly under the low irradiance (light intensity) conditions that afflict many artificially illuminated environments.

Another consideration that wearable solar cell systems must take into account is the potential harm to public health imposed by toxic materials associated with some PV materials. In traditional solar cell systems and installations, toxicity of materials comes into play only when solar panels reach the end of their useful lifetime and must be disposed of and recycled. In wearables and portables, however, toxic materials can potentially pose a risk during normal usage or during breakage and cracking which come about as part of normal everyday usage. The toxicity of cadmium (Cd) in CdTe solar cells and cadmium and lead (Pb) in many quantum dot and some PSCs often eliminates them from consideration in wearable products as well. DSCs are often made with ruthenium dyes which are highly toxic and carcinogenic. Given the propensity of DSCs made with liquid electrolyte to leak, the toxicity of ruthenium may be even more concerning than toxic materials in solid PV cells. To a lesser degree, CIGS solar cells contain smaller amounts of cadmium which might be seen as more acceptable. However, CIGS is ill suited to wearable solar cell systems for other reasons, particularly poor performance in low light conditions.

Many of the third-generation PV cells in Tables 9.1 through 9.3 also have problems with stability which drastically reduces their useful lifetime and makes them impractical for commercialization. For example, the longest reported lifetime for PSCs to date is about 1 year, which compares very poorly with the standard 25 year lifetime associated with silicon solar panels (Meng, You, and Yang 2018). Many OPV cells last less than 1 year outdoors and a lifetime of greater than 2 years has not yet been demonstrated for any OPV technology under any conditions (Zhang et al. 2018). Dye-sensitized PV cell lifetimes are limited by leakage of liquid electrolyte from the cell itself and certain types of quantum dot solar cells decompose when exposed to air (Wang et al. 2014).

TABLE 9.3

Compatibility with Wearable Solar Cell Systems

PV Technology	Flexible	Low Cost	Low Light[a]	Toxicity	Tunable
Monocrystalline Si	No	No	Worse	Low	No
Polycrystalline Si	No	No	Worse	Low	No
GaAs	Yes	No	Similar	Moderate	No
Amorphous Si (thin film)	Yes	Yes	Similar	Low	No
CIGS (thin film)	Yes	No	Worse	Moderate	Yes
CdTe (thin film)	Yes	Yes	Similar	High	No
Organic	Yes	Yes	Better	Somewhat	Yes
Dye-sensitized	Yes	Yes	Better	High[b]	No[c]
Perovskite	Yes	Yes	Better	High[d]	Yes
Quantum dot	Yes	Yes	Similar	High[e]	Yes

[a] PV cell performs better, worse, or similar at low light compared to exposure to sunlight at 1,000 W/m².
[b] Many dye-sensitized solar cells contain ruthenium which is highly toxic and carcinogenic.
[c] DSC PV cells demonstrate tunable bandgaps when designed on a quantum dot scale.
[d] Many perovskites contain lead (Pb) which is carcinogenic and causes neurological damage.
[e] Many quantum dot PV cells contain cadmium or lead.

9.3 Array Considerations

The good news about wearable solar cell systems is that their design often requires that fewer PV cells be connected in series in the construction of PV arrays and fewer PV cells be used overall in such arrays. Smaller numbers of PV cells come about because the energy demand of wearable systems is significantly lower than traditional systems. And, because most wearables and portables operate on a power supply voltage that is compatible with standard 5-V USB-based charging, the difference between the output voltage of a single solar cell and the desired input voltage to the devices relying on those PV cells for power is much smaller than in traditional systems. Since even basic boost converter designs can provide five times the input voltage at the output voltage with reasonable efficiency (Erickson 2007), arrays that contain strings of only one or two cells in series are possible.

In a more traditional solar cell system with much longer (series-connected) strings of PV cells, one malfunctioning or underperforming cell can create a potential hot spot, which at best, sinks much of the power generated by the string and at worst, damages the cell through excessive heat dissipation. To avoid hot spots, underperforming cells are often bypassed and the power generated by the offending PV cell is lost in favor of preserving the integrity of the array. Losing a cell in a string can decrease the voltage in that string which decreases the voltage of the array and results in a total power that is less than the maximum power potential of all the cells in the array. Strategies to deal with underperforming cells in dense arrays are often complex and consume non-trivial amounts of power during reconfiguration of arrays or other array management strategies designed to accommodate underperformers. Limiting the number of cells connected in series drastically reduces the complexity of array management and reduces power losses caused by these underperformers.

In summary, the design of PV arrays for wearable solar cell systems must balance the penalty imposed by an underperforming cell with efficiency losses incurred by array designs which minimize this penalty. For a large number of wearables and portables (excluding laptop computers which typically require much higher voltages to operate), eliminating strings altogether or reducing string size to two or three cells strikes this balance. The number of strings connected in parallel to complete the array is then a function primarily of the total energy demand of the solar energy system.

9.4 Charge Controller and Battery Considerations

Lightweight, lithium-based batteries have the highest energy density and specific energy of any mainstream batteries. For this reason, they are likely to remain the rechargeable battery of choice for any portable or wearable that consistently consumes more than a few milliwatt-hours (mWh). And, because they are suitable for portable power banks, lithium-based batteries can also supplement or displace single use batteries even in very-low-power wearables and portables. In an ideal world, a single lithium-based battery pack would power an entire plethora of wearable and portable devices worn or carried by an individual through a seamless network of (wired, wireless, or both) charging links that extend from head to toe.

Wonderful as they are, however, lithium-based batteries have some idiosyncrasies with respect to how they are charged. As discussed in Chapter 7, lithium batteries are charged using a constant current-constant voltage approach with some modifications. During the first stage of charging, a current around 0.5C mA (where C is the capacity of the battery in mAh) supplies energy to the battery. Undercharging at lower currents compromises battery capacity while overcharging can overheat the battery and compromise safety. Once a certain voltage is reached, the battery must be charged at a constant voltage (and decreasing current) to prevent overcharging. Overcharging occurs when all ions in the battery have finished moving during the charging cycle and results in overheating and potential explosion due to outgassing in the battery electrolyte (Digikey 2016; Battery University 2018).

To avoid damaging or compromising the lifetime of a lithium-based battery, both voltage across and current through the battery must be controlled. Unfortunately, wearable solar cell systems have a terrible track record for providing both constant current and constant voltage. Exposure to light source intensities that vary by several orders of magnitude and complex, dynamic, and frequently changing shading patterns on a wearable system spell doom for a lithium-based battery. Connecting an array of wearable solar cells directly to a lithium-based battery would not only be inefficient and destructive, but also potentially dangerous.

But, all is not lost. Constant voltage conditions can be maintained by using a charge controller to adjust the duty cycle of a DC-DC converter connected between the solar cell array and the battery. To some extent, current can be limited (i.e., reduced) by using shunt or series regulation (Chapter 7), but this approach tends to waste current or allow some undercharging of the battery. This leads to the need for some kind of temporary energy storage in addition to the battery itself to regulate the current without a major diversion or loss of power. A supercapacitor is one such temporary storage device that, unlike a lithium-based battery, will take a much broader range of charging currents. Once the supercapacitor has stored sufficient energy to supply the battery with a constant current, it can reconnect to the charging cycle, supplying the desired current until such time that it can't, and then disconnect from the battery to recharge once again. Given the nature of recharging demand in wearables and portables, however, it is conceivable that battery-based charging could be reduced or eliminated altogether if temporary energy storage technology, such as that provided by supercapacitors, continues to advance.

In summary, wearable solar cell systems require a DC-DC converter not only to boost voltage but also to charge batteries appropriately. In addition, to maximize battery life (hours between charges) and battery lifetime (total number of recharging cycles before the battery is permanently discharged), a temporary storage device or other means to efficiently control current into the battery being charged should be designed into the overall solar cell system.

9.5 Surface Area Considerations

Wearable solar cell systems cannot possibly compete with traditional systems (like those installed on a rooftop) in terms of total power generated in a typical day. Among other reasons, the surface area of the human body is nowhere near that of a rooftop; the average residential roof in the United States has a surface area of about 241 m^2 (Center for Sustainable Systems n.d.) while the surface area of an average adult is less than 2 m^2 (Shiel 2018). Furthermore, in the everyday activity of most individuals, wearable PV will not be exposed to nearly as much total light energy as rooftop panels. For example, an individual working in an office under fluorescent lighting will be exposed to only between 6.4 and 9.1 W/m^2. In contrast, the average rooftop exposed to full sunlight on a clear day receives approximately 82.7 W/m^2 (Table 9.4).

TABLE 9.4

Light Levels Associated with Common Light Sources

Condition	Illumination (lux)	Light Source Type	Conversion Factor (lm/W)	Conversion Factor (W/lm)	Illumination (W/m²)
Natural Light					
Direct sunlight (clear day)	107,527	Sun	70–105	0.0096–0.0143	1,024–1,536
Full daylight (clear day)	10,752		130	0.0077	82.7
Full daylight (overcast)	1,075		110	0.0091	9.77
Night (full moon)	0.108	Moon	130	0.0077	0.0008
Night (overcast)	0.0001		110	0.0091	Negligible
Outdoor Artificial Lighting					
Highways	4.0–14.0	Metal halide	65–116	0.0086–0.0154	0.035–0.215
Major roadways	6.0–17.0				0.052–0.262
Local roadways	3.0–9.0				0.026–0.139
Indoor Artificial Lighting					
Classrooms	150	Fluorescent	55–78	0.0128–0.0182	1.92–2.73
		Incandescent	14–18	0.0556–0.0714	8.33–10.7
		LED	55–93	0.0108–0.0182	1.61–2.73
Home	250	Fluorescent	55–78	0.0128–0.0182	3.21–4.55
		Incandescent	14–18	0.0556–0.0714	13.9–17.9
		LED	55–93	0.0108–0.0182	2.69–4.55
Office	500	Fluorescent	55–78	0.0128–0.0182	6.41–9.09
		Incandescent	14–18	0.0556–0.0714	27.8–35.7
		LED	55–93	0.0108–0.0182	5.38–9.09
Workspace (difficult visual tasks)	5,000–10,000	Fluorescent	55–78	0.0128–0.0182	64.1–182
		Incandescent	14–18	0.0556–0.0714	278–714
		LED	55–93	0.0108–0.0182	53.8–182

Source: National Optical Astronomy Observatory (n.d.), Littlefair (1985), Choudhury (2014), and Ledke Technology Ltd. (n.d.).

What can 9.1 W/m² achieve? The answer to that question certainly depends on the technology and method by which PV cells are made wearable for the average consumer. Consider first the use of highly efficient but rigid crystalline silicon with 26.1% power conversion efficiency in ideal AM 1.5 conditions (Table 9.1). Because of a lack of flexibility, the rigid panel might be designed to be worn across the outward-facing side of a backpack with a surface area of about 0.1 m². Under fluorescent lighting, silicon is only 34% as efficient as it is when exposed to sunlight (Table 9.2). Thus, for every watt of light reaching the panel, a maximum of 0.09 W is successfully harvested (34% of the 26.1% power conversion efficiency in response to sunlight). 9.1 W/m² over 0.1 m² of area amounts to 0.91 W and at 9% efficiency, yields 0.082 W, which over 16 hours of exposure to artificial fluorescent light, yields 1.3 Wh of energy. Compared to the per capita electricity consumption per day in the United States (U.S. Energy Information Administration 2017), this seems like a trivial amount of energy. However, compared to the energy demand of wearables and portables (Tables 8.1 through 8.5), 1.3 Wh of energy is enough to support an MP3 player, a wireless headset, a smartwatch, a fitness band, a sports watch for many hours, or a hearing aid for multiple days,. And, if the office worker goes outside for lunch, the 0.1 m² backpack panel will enjoy an additional 82.7 W/m² (Table 9.4) for an hour, adding 2.2 Wh to the energy harvested bringing the total to approximately 3.5 Wh over a 16-hour day, enough to power a laptop or tablet computer for an hour or two or a smartphone for most of the day. Leaving the backpack on a sunny window ledge can increase the total energy harvested even further.

Another viable, wearable solar cell system scenario involves the use of more flexible PV cells integrated into clothing as well as other wearable accessories. Flexible PV cells tend to be less efficient than more rigid crystalline silicon-based PV cells but because of their flexibility, can cover irregular topologies, thus making them suitable for integration into clothing. However, the flexibility of PV cells on plastic substrates is often limited to bending in only one direction and most flexible PV cells therefore lack the full conformal flexibility of textile-based clothing (Schubert and Werner 2006). For this reason, the placement of these cells onto clothing is limited to areas on the body which usually flex only unidirectionally like the back and shoulders. Consider a scenario where flexible OPVs with a maximum efficiency of 15.6% in sunlight (Table 9.1) and 11.2% efficiency in fluorescent light (Table 9.2) are integrated onto clothing on the upper back, shoulders, and chest which account for 9%, 2%, and 9% of the body's surface area, respectively (Michigan Medicine n.d.). Combined with OPVs integrated onto a hat that covers most of the head and neck, equivalent to another 9% of the body's surface area, the total PV surface area for an average adult with a total of 1.7-m² body surface area (Shiel 2018) would be about 0.49 m². For an 8-hour day in the office, clothing and hat would collect approximately 4.0 Wh of energy under fluorescent light conditions over the work day, increasing to 10.3 Wh if an hour outdoors is included in the estimate. In this example, the added flexibility of the OPV shows the potential to overcome its reduced efficiency compared to silicon via integration onto a larger surface area on the body.

A third scenario involves providing full conformal flexibility to wearable solar cell clothing by integrating PV cells into textiles so that they can be woven into clothing and cover most of the body surface area. Admittedly, such an approach to wearable solar cells is further out into the future than rigid silicon panels or flexible organic PV panels as described in the previous two scenarios. Flexible PV cells rely on a continuous planar substrate in two dimensions that allow the PV cell to bend but not crinkle. Crinkling involves giving up on this continuous substrate and allowing fibers to move against one another which can create a myriad of design challenges including moving interconnects, short circuits, loss of signal, prohibitive shading and other problems (Schubert and Werner 2006). Despite these challenges, research continues to advance toward truly conformal, wearable solar clothing (Weng et al. 2016). Fiber-shaped solar cells that can be woven into PV textiles have been demonstrated with DSCs using TiO_2 and carbon/graphene nanotubes at an energy conversion efficiency of 8.4% (Sun et al. 2014), with OPVs that avoid the potential for leakage inherent in most DSCs at 3.8% efficiency (Lee et al. 2009), and with PSCs at 3.3% efficiency (Qiu et al. 2014). PV cells have also been made directly from textile substrates using polymer PV materials at 1.8% efficiency (Lee et al. 2014), and with a DSC on cotton fabric (Sahito et al. 2015). In their current state-of-the-art, PV-based fibers and textiles offer significantly increased surface area for collecting light but at maximum efficiencies that are about 30% of the efficiency of silicon-based solar panels.

9.6 Summary

Flexible, wearable clothing capable of converting ambient light to enough power to support mobile energy needs is no longer an invention of science fiction. Rigid, silicon-based solar panels to support energy needs for hiking and other outdoor sports are already commercially available (Goal Zero n.d.) and recent advances in flexible second- and third-generation solar cells which can be readily printed using a wide range of inexpensive fabrication techniques open the door to smart, wearable solar cell clothing of all kinds.

In terms of today's PV cell capability, the bottom line is that powering wearables with wearable solar cell systems is a realistic proposition for most individuals who spend some part of the day outdoors, work under typical office lighting conditions for some part of the day, and avoid diving into a deep dark cave after work. Those who spend more time outdoors or benefit from direct sunlight for a few hours of the day have the potential to power both wearables and portables seamlessly and without interruption of service. While wearable solar cell systems may not be the whole solution to powering all the portable and wearable devices individuals choose to carry on their person, they can definitely contribute to reducing the environmental impacts of these devices either directly through displacement of fossil fuels or through the reduced consumption of batteries. And, last but certainly not least, these wearable solar cell systems can open the door to an even broader selection of wearables, many of which can have profound, positive impacts on quality of life.

References

Battery University. 2018. "BU-409: Charging Lithium Ion." April 24, 2018. https://batteryuniversity.com/learn/article/charging_lithium_ion_batteries.

Center for Sustainable Systems. n.d. "Residential Buildings Factsheet." Accessed March 8, 2019. http://css.umich.edu/factsheets/residential-buildings-factsheet.

Choudhury, Asim Kumar Roy. 2014. "Characteristics of Light Sources." In *Principles of colour and appearance measurement: Object appearance, colour perception and instrumental measurement*. Elsevier. http://dx.doi.org/10.1533/9780857099242.1.

Digikey. 2016. "A Designer's Guide to Lithium Ion (Li-Ion) Battery Charging." September 1, 2016. https://www.digikey.com/en/articles/techzone/2016/sep/a-designer-guide-fast-lithium-ion-battery-charging.

Erickson, Robert W. 2007. "DC–DC Power Converters." In *Wiley Encyclopedia of Electrical and Electronics Engineering*. John Wiley & Sons. https://doi.org/10.1002/047134608X.W5808.pub2.

Goal Zero. n.d. "Portable Solar Panels." Accessed July 19, 2019. https://www.goalzero.com/.

Ledke Technology Ltd. n.d. "What Is Luminous Efficacy?" Accessed July 9, 2019. http://www.ledke.com/what-is-luminous-efficacy-definition/.

Lee, Michael R., Robert D. Eckert, Karen Forberich, Gilles Dennler, Christoph J. Brabec, and Russell A. Gaudiana. 2009. "Solar Power Wires Based on Organic Photovoltaic Materials." *Science* 324 (5924): 232–235.

Lee, Seungwoo, Younghoon Lee, Jongjin Park, and Dukhyun Choi. 2014. "Stitchable Organic Photovoltaic Cells with Textile Electrodes." *Nano Energy* 9 (October): 88–93. https://doi.org/10.1016/j.nanoen.2014.06.017.

Littlefair, Paul J. 1985. "The Luminous Efficacy of Daylight: A Review." *Lighting Research & Technology* 17 (4): 162–182. https://doi.org/10.1177/147715358501700 40401.

Meng, Lei, Jingbi You, and Yang Yang. 2018. "Addressing the Stability Issue of Perovskite Solar Cells for Commercial Applications." *Nature Communications* 9 (1): 5265. https://doi.org/10.1038/s41467-018-07255-1.

Michigan Medicine. n.d. "Estimating the Size of a Burn." Accessed July 11, 2019. https://www.uofmhealth.org/health-library/sig254759.

Minnaert, Ben, and Peter Veelaert. 2014. "A Proposal for Typical Artificial Light Sources for the Characterization of Indoor Photovoltaic Applications." *Energies* 7 (3): 1500–1516. https://doi.org/10.3390/en7031500.

National Optical Astronomy Observatory, NOAO. n.d. "Recommended Light Levels (Illuminance) for Outdoor and Indoor Venues." https://www.noao.edu/education/QLTkit/ACTIVITY_Documents/Safety/LightLevels_outdoor+indoor.pdf.

NREL (National Renewable Energy Laboratory). 2019. "Best Research-Cell Efficiency." https://www.nrel.gov/pv/cell-efficiency.html.

Polman, Albert, Mark Knight, Erik C. Garnett, Bruno Ehrler, and Wim C. Sinke. 2016. "Photovoltaic Materials: Present Efficiencies and Future Challenges." *Science* 352 (6283): 1–10. https://doi.org/10.1126/science.aad4424.

Qiu, Longbin, Jue Deng, Xin Lu, Zhibin Yang, and Huisheng Peng. 2014. "Integrating Perovskite Solar Cells into a Flexible Fiber." *Angewandte Chemie International Edition* 53 (39): 10425–10428.

Sahito, Iftikhar Ali, Kyung Chul Sun, Alvira Ayoub Arbab, Muhammad Bilal Qadir, and Sung Hoon Jeong. 2015. "Graphene Coated Cotton Fabric as Textile Structured Counter Electrode for DSSC." *Electrochimica Acta* 173: 164–171.

Schubert, Markus B., and Jürgen H. Werner. 2006. "Flexible Solar Cells for Clothing." *Materials Today* 9 (6): 42–50.

Shiel, William C. 2018. "Definition of Body Surface Area." MedicineNet. https://www.medicinenet.com/script/main/art.asp?articlekey=39851.

Sun, Hao, Xiao You, Jue Deng, Xuli Chen, Zhibin Yang, Jing Ren, and Huisheng Peng. 2014. "Novel Graphene/Carbon Nanotube Composite Fibers for Efficient Wire-Shaped Miniature Energy Devices." *Advanced Materials* 26 (18): 2868–2873.

U.S. Energy Information Administration, EIA. 2017. "Per Capita Residential Electricity Sales in the U.S. Have Fallen since 2010." https://www.eia.gov/todayinenergy/detail.php?id=32212.

Wang, Ke, Weiwei He, Luo Wu, Guoping Xu, Shulin Ji, and Changhui Ye. 2014. "On the Stability of CdSe Quantum Dot-Sensitized Solar Cells." *RSC Advances* 4 (30): 15702–15708.

Weng, Wei, Peining Chen, Sisi He, Xuemei Sun, and Huisheng Peng. 2016. "Smart Electronic Textiles." *Angewandte Chemie International Edition* 55 (21): 6140–6169. https://doi.org/10.1002/anie.201507333.

Zhang, Yiwei, Ifor D. W. Samuel, Tao Wang, and David G. Lidzey. 2018. "Current Status of Outdoor Lifetime Testing of Organic Photovoltaics." *Advanced Science* 5 (8). https://doi.org/10.1002/advs.201800434.

Glossary

AC: alternating, sinusoidally varying current and voltage used to deliver electricity to homes, office buildings, etc.

back-surface field layer: a layer of material near the back-side of a solar cell that prevents free electrons and holes from recombining.

bandgap: the difference in energy levels between the highest valence band energy and the lowest conduction band energy in a material.

battery life: the length of time a device can run on a full charge of a battery.

battery life span: see battery lifetime.

battery lifetime: the average number of recharging cycles that a rechargeable battery provides during normal operation before becoming unusable.

conduction band: the energy levels that excited electrons can occupy in a semiconductor or insulator.

DC: direct, constant current and voltage used to power many electronic devices that operate on low voltages or are portable.

degenerate semiconductor: a semiconductor that is so heavily doped that it acts more like a metal than a semiconductor with the Fermi level situated very near to or in the valence or conduction band.

efficiency: usually refers to the energy conversion efficiency of a solar cell.

electron-hole pair: refers to an electron and hole generated by the absorption of a photon in materials such as inorganic semiconductors where the electron and hole are only loosely bound to one another with an energy on the order of meV.

energy conversion efficiency: the ratio of output electrical power to input light power in a solar cell.

energy demand (with reference to wearable solar cell systems): the average power consumed by an electronic device or appliance over the course of its battery life.

exciton: refers to an electron and a hole generated by the absorption of a photon in materials such as organic semiconductors where the electron and the hole remain bound to one another with a significantly greater binding energy than an electron-hole pair.

external quantum efficiency: the ratio of current or charge carriers collected by a solar cell to the photons that are incident upon the cell.

Fermi level: the energy level associated with a 50% chance or probability of being occupied by a free electron or hole in a semiconductor; the energy at which 50% of free current carriers (electrons or holes) have greater energy and 50% have lower energy than the Fermi level.

fill factor: the degree to which the voltage versus current curve in a solar cell approaches the ideal solar cell characteristic; the ratio of the maximum power produced by the cell to the product of the open circuit voltage and short circuit current.

HOMO: the energy associated with the highest occupied molecular orbital in a material which is roughly analogous to the valence band edge in an inorganic semiconductor.

hot spot: a condition in an array of solar cells or solar panel whereby an underperforming cell goes into large reverse bias, steals power from the array rather than supplying it, and experiences substantial temperature increases that can lead to cracking and permanent damage.

insulator: a non-conductive material characterized by conduction and valence bands that are separated by a large bandgap.

internal quantum efficiency: the ratio of electrons excited to a higher energy level in a solar (photovoltaic) cell to the number of photons absorbed by the cell.

irradiance: the amount of light striking (i.e., incident upon) a surface (such as that of a solar cell) per unit area, as measured in radiometric units such as W/m^2.

light intensity: an ambiguous term that most frequently refers to irradiance or the total light (in watts) incident on a device such as a solar cell.

LUMO: the energy associated with the lowest unoccupied molecular orbital in a material that is roughly analogous to the conduction band edge in an inorganic semiconductor.

metal: a conductive material characterized by conduction and valence bands that overlap.

nondegenerate semiconductor: a semiconductor that is not heavily doped, such that the Fermi level is at least $3kT$ (where k is the Boltzmann constant and T is temperature in K) above the valence band or $3kT$ below the conduction band.

open circuit voltage: the voltage across a photovoltaic cell when the cell draws no (zero) current.

partial shading: a condition in an array of solar cells or a solar panel where one or more cells experience significantly lower light energy than other cells; often referred to simply as shading.

photoelectric effect: the phenomenon that enables electrons to absorb light energy and move to the vacuum level.

photovoltaic array: an array of photovoltaic cells consisting of strings connected in parallel to one another.

photovoltaic cell: a device that converts light energy to electrical energy through the photovoltaic effect.

photovoltaic effect: the phenomenon that enables electrons to absorb light energy and move to higher energy levels; these higher energy levels are restricted to energies less than that of the vacuum level.

photovoltaic string: photovoltaic cells connected in series.

power conversion efficiency: see energy conversion efficiency.

recharging lifetime: see battery lifetime.

semiconductor: a material in which the conduction and valence bands are separated by a bandgap that is significantly smaller than that of an insulator.

soiling: a condition by which a solar cell is covered with dirt, dust, or other material, which results in significantly less power generated for the same light intensity as an unsoiled but otherwise identical solar cell.

solar cell: a less formal term used to describe a photovoltaic cell.

solar cell system: a system consisting of solar cells arranged in arrays with array management and support electronics designed to convert light energy to electricity and supply energy to energy storage devices or drive AC or DC devices and appliances.

vacuum level: the energy of an electron in free space (i.e., vacuum) where it is no longer associated with any atom, molecule, or material.

valence band: the energy levels that electrons in their resting states can occupy in a material.

wavelength: a property of light defined by the distance between two adjacent and identical points in a light wave (e.g., two peaks).

wavelength spectrum (or spectrum): the combination of wavelengths and their intensities that define the color of a light source.

window layer: a layer of material near the surface of a solar cell that prevents electrons and holes from recombining at the surface.

Index